TABLE OF CONTENT

Chapter 1: Introduction to the Amazon Rainforest 4
    The Amazon's Enigmatic Landscape: A Global Treasure 4
    A Tapestry of Biodiversity: Exploring the Richness of Life 7
    Understanding the Ecosystem: A Complex Interplay of Forces 8
    The Importance of Conservation: Protecting a Precious Heritage 10

Chapter 2: The Canopy: Life Among the Branches 12
    A World of Light and Air: Exploring the Canopy's Unique Environment 12
    The Canopy's Inhabitants: Adaptations for Survival in the Heights 13
    The Canopy's Role in the Ecosystem: A Vital Bridge for Life 19
    Exploring the Canopy: Methods of Research and Exploration 22

Chapter 3: The Understory: A World of Shadows and Shade 25
    The Understory's Environment: A Realm of Constant Change 25
    Adaptation and Survival in the Understory: A Challenge for Life 27
    Key Players in the Understory: Animals and Plants of the Shade 28
    The Understory's Importance: A Nursery for Life 34

Chapter 4: The Forest Floor: Life in the Dark 36
    The Forest Floor's Environment: A World of Decomposition and Rebirth 36
    Life on the Forest Floor: Adaptations for Low-Light and Moisture 37
    Decomposers and their Role: Breaking Down and Recycling Nutrients 39
    The Forest Floor's Importance: A Foundation for Life 42

Chapter 5: The Rivers: Veins of Life 44
    The Amazon River System: A Network of Life and Movement 44
    The River's Impact on the Forest: A Constant Source of Change 44
    The River's Inhabitants: Adaptations for Aquatic Life 47
    The River's Importance: A Pathway for Nutrients and Life 48

Chapter 6: The Trees: Pillars of the Forest 51
    The Diversity of Amazonian Trees: A Wide Range of Adaptations 51
    The Tree's Role in the Ecosystem: Providing Structure and Shelter 52
    The Importance of Trees in the Carbon Cycle: Regulating Climate 54
    Threats to Amazonian Trees: Deforestation and Climate Change 56

Chapter 7: The Birds: Symphony of the Rainforest 59

The Amazon's Avian Diversity: A Colorful Spectacle 59
Adaptations for Flight and Survival: A World of Specializations 60
The Birds' Role in the Ecosystem: Seed Dispersal and Pollination 62
Birds in Peril: Threats to Amazonian Bird Populations 65

Chapter 8: The Mammals: Giants and Shadows 67
The Amazon's Mammalian Diversity: A Range of Shapes and Sizes 67
Adaptations for Survival in the Jungle: From Stealth to Strength 68
The Mammals' Role in the Ecosystem: Predators, Herbivores, and More 70
Threats to Amazonian Mammals: Habitat Loss and Hunting 72

Chapter 9: The Reptiles and Amphibians: Masters of Camouflage and Survival 74
The Amazon's Reptile and Amphibian Diversity: A World of Camouflage and Poison 74
Adaptations for Survival: Camouflage, Venom, and Specialized Defenses 75
The Reptiles and Amphibians' Role in the Ecosystem: Predators, Prey, and More 77
Threats to Amazonian Reptiles and Amphibians: Habitat Loss and Climate Change 79

Chapter 10: The Insects: A Buzzing World of Life 82
The Amazon's Insect Diversity: A Microcosm of Life 82
The Insects' Role in the Ecosystem: Pollination, Decomposition, and Food Chains 84
The Importance of Insects: A Keystone to Life in the Jungle 85
Threats to Amazonian Insects: Habitat Loss and Pesticides 87

Chapter 11: The Fungi: Masters of Decomposition 89
The Amazon's Fungi Diversity: A Hidden Realm of Life 89
The Fungi's Role in the Ecosystem: Breaking Down Dead Matter and Recycling Nutrients ... 90
The Importance of Fungi: Supporting Life in the Jungle 93
Threats to Amazonian Fungi: Habitat Loss and Pollution 94

Chapter 12: The Plants: A Symphony of Colors and Textures 97
The Amazon's Plant Diversity: A World of Beauty and Adaptability 97
Adaptations for Survival in the Jungle: From Photosynthesis to Defense Mechanisms 99
The Plants' Role in the Ecosystem: Producing Oxygen, Providing Food, and Supporting Life
.... 100
Threats to Amazonian Plants: Habitat Loss and Climate Change 102

Chapter 13: The Future of the Amazon: Challenges and Opportunities 105
The Amazon's Importance to the World: A Global Resource and a Vital Ecosystem 105
Threats to the Amazon: Deforestation, Climate Change, and Pollution 106

The Need for Conservation: Protecting a Precious Heritage 108
Opportunities for Sustainability: Working Towards a Balanced Future 110

Chapter 1: Introduction to the Amazon Rainforest

The Amazon's Enigmatic Landscape: A Global Treasure

The Amazon's Enigmatic Landscape: A Global Treasure.

The Amazon rainforest, a sprawling emerald canvas draped across the heart of South America, is more than just a verdant expanse. It is a living testament to the Earth's enduring power, a complex tapestry woven with vibrant life and intricate ecosystems, a treasure trove of biodiversity unmatched anywhere else on the planet. To truly grasp the magnitude of its significance, one must transcend the superficial beauty of its towering canopy and delve into the depths of its enigmatic landscape.

The Amazon is a realm of extremes, a symphony of contrasting forces. Sunlight filters through the dense foliage, dappling the forest floor in a mosaic of light and shadow. A constant hum of life permeates the air, a chorus of insects, birds, and mammals weaving their unique melodies into the intricate rhythm of the rainforest. The air itself, thick with moisture and the scent of decaying vegetation, evokes a sense of both vitality and decay, a testament to the relentless cycle of life and death that drives the Amazon's perpetual motion. .

This symphony of life is made possible by the intricate web of interactions that govern the Amazon's ecosystem. From the towering trees that act as guardians of the forest, their roots anchoring the soil and their leaves forming a verdant canopy that shelters a myriad of creatures below, to the smallest insects, diligently playing their part in the complex processes of decomposition and pollination, every organism contributes to the intricate dance of life that defines the Amazon.

The Amazon's landscape is a mosaic of diverse habitats, each teeming with life uniquely adapted to its specific environment. Along the vast network of rivers that snake their way through the rainforest, life thrives in a perpetual state of flux. Floodplains, inundated during the rainy season, transform into vibrant meadows teeming with aquatic life, while the riverbanks, perpetually exposed to the whims of the currents, are home to a hardy array of species that have learned to adapt to the ever-changing environment.

The Amazon's rivers are more than just waterways; they are arteries of life, carrying nutrients and sediment throughout the rainforest, connecting diverse ecosystems and ensuring the flow of energy through the vast expanse. As the mighty Amazon River meanders its way to the Atlantic Ocean, it becomes a conduit for countless species, transporting them from the depths of the rainforest to the bustling shores of the ocean, and vice versa.

The Amazon's landscape is characterized by an astonishing diversity of flora, with an estimated 40,000 plant species, a testament to the rainforest's ability to sustain life in its myriad forms. From the majestic kapok trees that tower over the forest canopy, their massive trunks a symbol of resilience and strength, to the delicate orchids that adorn the forest floor, their vibrant colors a testament to the Amazon's aesthetic brilliance, the rainforest's flora is a kaleidoscope of life, each species playing its own role in the complex web of interactions that defines the ecosystem.

The Amazon's flora is not only aesthetically stunning; it is also a testament to the resilience of life. The rainforest's ability to withstand the onslaught of natural disasters, from floods to droughts, is a testament to the inherent adaptability of its flora and fauna. Species have evolved to survive in these challenging conditions, developing unique strategies to thrive in the face of adversity. This remarkable resilience is a testament to the Amazon's ability to adapt and endure, even in the face of environmental pressures.

The Amazon's flora is a vital resource for the indigenous communities who have called the rainforest home for millennia. These communities have developed a profound understanding of the rainforest's resources, using its plants for food, medicine, and building materials. Their knowledge, passed down through generations, is a testament to the deep connection between humans and the natural world, and a vital source of information for modern scientists seeking to unravel the mysteries of the rainforest's vast biodiversity.

The Amazon's fauna is just as impressive as its flora, boasting an astounding array of species, estimated to encompass 10% of the world's known species. From the majestic jaguars, apex predators that roam the rainforest, their sleek coats a testament to their power and grace, to the tiny hummingbirds, their wings a blur as they flit through the forest canopy, feeding on nectar from the vibrant flowers, the Amazon's fauna is a living testament to the diversity of life on Earth.

The Amazon's fauna is a testament to the interconnectedness of all living things. Predators and prey, parasites and hosts, all play their part in the intricate dance of life that defines the rainforest's ecosystem. The balance between these interactions is delicate, and any disruption can have cascading effects throughout the entire ecosystem. This delicate balance underscores the importance of preserving the Amazon's natural integrity, for the sake of the countless species that call it home, and for the well-being of the planet as a whole.

The Amazon's fauna is a vital source of inspiration for scientists and artists alike. The rainforest's diverse array of species, each with its unique adaptations and behaviors, provides a wealth of knowledge for researchers seeking to understand the intricacies of life on Earth. The Amazon's beauty, both visually and ecologically, has inspired artists for generations, from the paintings of the Impressionists to the photography of modern naturalists. The Amazon's aesthetic and ecological richness is a testament to the enduring power of nature, a source of both inspiration and wonder.

The Amazon is not just a vast expanse of rainforest; it is a complex and dynamic ecosystem, a living laboratory that embodies the beauty and fragility of life on Earth. Its vast biodiversity, its intricate web of interactions, and its resilience in the face of adversity, all speak to the power and importance of this global treasure. The Amazon is a testament to the enduring power of nature, a reminder that even in the face of human encroachment, life persists, adapting and evolving to navigate the challenges of an ever-changing world.

The Amazon's future is inextricably linked to the fate of humanity. As we continue to expand our footprint on the planet, it is imperative that we recognize the importance of preserving this vital ecosystem, not only for the sake of the countless species that call it home, but for the well-being of the entire planet. The Amazon is a global treasure, a reminder that we are but one part of a complex and interconnected web of life, and that our actions have profound consequences for the health and well-being of the entire planet.

The Amazon's story is one of resilience, adaptation, and wonder. As we delve deeper into its enigmatic landscape, we discover a world of beauty and complexity, a testament to the enduring power of nature and the interconnectedness of all living things. The Amazon is a treasure that must be protected, for the sake of the planet and for the future of humanity. It is a legacy that we must safeguard, ensuring that the vibrant symphony of life that defines the Amazon will continue to resonate for generations to come. .

A Tapestry of Biodiversity: Exploring the Richness of Life

A Tapestry of Biodiversity: Exploring the Richness of Life.

The Amazon rainforest, a verdant expanse spanning nine nations, pulsates with life. It is a symphony of vibrant colors, intoxicating scents, and a cacophony of sounds. This symphony, however, is more than just a spectacle; it is a testament to the intricate tapestry of biodiversity that blankets the region. The Amazon, a crucible of evolution for millennia, has sculpted a breathtaking array of life forms, each intricately woven into the fabric of this ecosystem. .

To truly appreciate the Amazon's biological wealth, one must delve into the intricacies of its flora and fauna. A seemingly endless parade of plant species, from towering emergent trees like the Brazil nut, whose fruits provide sustenance for countless creatures, to the delicate orchids clinging to their branches, each plays a crucial role in maintaining the rainforest's delicate balance. The understory is equally vibrant, teeming with a kaleidoscope of herbs, ferns, and mosses, each competing for light and nutrients in this densely packed habitat. These plants, in turn, support a bewildering array of animals, from the majestic jaguar, the apex predator of the jungle, to the minuscule hummingbirds, their iridescent plumage a shimmering testament to nature's artistry.

The Amazon's biodiversity is not just a matter of numbers; it is a testament to the extraordinary adaptations that life has evolved to thrive in this unique environment. The caiman, its leathery hide blending seamlessly with the murky waters, patiently awaits its prey, while the spider monkey, with its prehensile tail and nimble fingers, navigates the canopy with ease. The vibrant macaws, their plumage a riot of colors, are not just beautiful to behold; their role as seed dispersers is vital for the rainforest's regeneration. Each species, from the smallest insect to the largest mammal, is a thread in this intricate tapestry, contributing to the health and resilience of the entire ecosystem.

The sheer diversity of the Amazon is a testament to the power of evolution, a process that has sculpted life into its current form over millions of years. This diversity, however, is not static. The rainforest is a dynamic entity, constantly evolving and adapting to the changing environment. The interactions between species, the intricate dance of predator and prey, the competition for resources, all contribute to the ever-changing mosaic of life.

But beyond its biological richness, the Amazon is a treasure trove of cultural diversity. Indigenous communities, for generations, have lived in harmony with the forest, their lives intricately woven into its rhythms. They possess an intimate understanding of the rainforest's plants, animals, and natural cycles, knowledge accumulated through centuries of observation and experience. Their traditional practices, their myths and legends, all speak to the deep connection between humanity and nature, a connection that is vital for the preservation of the Amazon's biological and cultural heritage.

As we delve deeper into the Amazon's tapestry of life, we uncover a world of complexity and beauty. The rainforest is not just a collection of flora and fauna; it is a vibrant, interconnected ecosystem, a testament to the enduring power of nature. Understanding the intricate web of life within the Amazon is not just a scientific endeavor; it is a vital step towards safeguarding this precious resource for future generations.

Understanding the Ecosystem: A Complex Interplay of Forces

Understanding the Ecosystem: A Complex Interplay of Forces.

The Amazon rainforest, a sprawling expanse of verdant life, pulsates with a symphony of interconnected forces. To comprehend its complexity, we must delve into the intricate tapestry of its ecosystem, a masterpiece woven by the delicate threads of climate, geology, biodiversity, and the relentless dance of evolution. The Amazon is not merely a collection of trees and animals; it is a living organism, a vast and dynamic system perpetually in flux, fueled by the energy of the sun and the relentless rhythm of the planet's cycles.

The very air we breathe plays a crucial role in shaping this ecosystem. The Amazon, often called the "lungs of the planet," generates a significant portion of Earth's oxygen. This oxygen production, a byproduct of photosynthesis, fuels the intricate web of life within the rainforest. However, the Amazon's role in regulating the Earth's climate extends far beyond oxygen production. Its vast canopy, a living filter, absorbs carbon dioxide from the atmosphere, mitigating the effects of climate change. This delicate balance between oxygen production and carbon sequestration underscores the Amazon's vital contribution to global climate stability.

Geological forces, too, have sculpted the Amazon's landscape, contributing to its unique biodiversity. The Amazon Basin, formed over millions of years by the uplift of the Andes Mountains, provides a vast network of rivers and tributaries that crisscross the rainforest. These waterways, lifeblood of the ecosystem, facilitate the dispersal of species, carry nutrients, and regulate water levels, creating a dynamic landscape. The Amazon's rich alluvial soils, nourished by the annual flooding, provide fertile ground for the rainforest's remarkable biodiversity. This interplay between geological forces and the dynamic river system forms the very foundation of the Amazonian ecosystem.

Life within the Amazon, a spectacular tapestry of diversity, thrives on these interconnected forces. From the towering canopy trees, home to an incredible array of epiphytes and arboreal creatures, to the murky depths of the river, teeming with fish and invertebrates, every species plays a crucial role in the intricate web of life. The rainforest's biodiversity is not merely a collection of fascinating creatures; it is a testament to the remarkable power of adaptation and co-evolution. The interplay of species, through relationships of predation, competition, and symbiosis, has sculpted the Amazon's biodiversity into a finely tuned symphony of life.

The Amazon, however, is not immune to change. The very forces that have shaped its ecosystem, from the cyclical rhythms of rainfall to the relentless march of human encroachment, pose both challenges and opportunities. The rainforest's resilience, forged over millennia of evolution, is tested by these pressures, prompting an ongoing dance of adaptation and change. Understanding the complexities of this ecosystem, the delicate balance of its forces, is crucial to ensuring its future. Conservation efforts, informed by scientific understanding, are paramount in preserving this invaluable natural treasure.

The Amazon, a microcosm of the planet's intricate interconnectedness, offers a powerful lesson in the delicate balance of nature. As we explore its complexities, we gain insights not just into the workings of this unique ecosystem, but into the interconnectedness of all life on Earth. The Amazon's story is a cautionary tale, a reminder of our responsibility to understand and protect the fragile ecosystems that sustain us. .

The Importance of Conservation: Protecting a Precious Heritage

The Amazon rainforest, a leviathan of emerald green sprawling across the heart of South America, is a symphony of life, a testament to the planet's enduring power and the intricate tapestry of biodiversity that graces our world. It stands as a beacon of hope, a vital ecosystem that sustains not only a vast array of flora and fauna but also the delicate balance of Earth's climate, a natural wonder that demands our respect and urgent protection.

The very air we breathe is intricately connected to the Amazon's well-being. Its towering trees act as colossal lungs, absorbing carbon dioxide and releasing oxygen, playing a crucial role in mitigating the escalating effects of climate change. The Amazon's vast canopy, a living tapestry of leaves and branches, intercepts sunlight, moderates temperatures, and regulates precipitation patterns, influencing weather systems across continents. Its intricate web of rivers and wetlands acts as a vital water reservoir, influencing global water cycles and ensuring the sustenance of countless communities reliant on its resources.

However, this magnificent ecosystem, a beacon of life, faces an unprecedented threat, a looming shadow cast by human activities that jeopardize its very existence. The relentless march of deforestation, driven by agricultural expansion, illegal logging, and resource extraction, has carved deep wounds into the Amazon's emerald heart. This relentless assault, fueled by unsustainable practices, not only diminishes the rainforest's ability to absorb carbon dioxide but also triggers a cascade of detrimental effects, pushing the ecosystem towards a tipping point.

The consequences of this ecological decline extend far beyond the Amazon's boundaries, reaching across continents and impacting global climate patterns. As deforestation progresses, the Amazon's capacity to moderate temperatures weakens, contributing to rising global temperatures and exacerbating the devastating effects of climate change. Reduced rainfall, a consequence of deforestation, can lead to drought, impacting agricultural production and exacerbating water scarcity, threatening the livelihoods of millions who depend on the rainforest's resources.

The Amazon's biodiversity, a vibrant tapestry woven with millions of species, is also under siege. The loss of habitat, driven by deforestation and fragmentation, threatens the survival of countless plant and animal species, many of which are endemic to the region and found nowhere else on Earth. As the rainforest's interconnected web of life unravels, the very fabric of the ecosystem weakens, leaving it vulnerable to cascading effects that could lead to irreversible consequences.

The importance of conservation in the Amazon rainforest cannot be overstated. It is not simply a matter of safeguarding a beautiful and awe-inspiring ecosystem but a crucial step in ensuring the health and well-being of our planet and the future of humanity. Protecting the Amazon is not just about preserving nature's beauty; it is about safeguarding the very foundation upon which life on Earth depends.

The task before us is daunting but not insurmountable. We must recognize the urgent need to shift from unsustainable practices to a sustainable model of living in harmony with the Amazon. This necessitates a concerted effort, encompassing a range of strategies aimed at reducing deforestation, promoting sustainable land management, fostering responsible resource extraction, and supporting conservation initiatives. .

International collaboration, policy changes, and community engagement are essential elements in this endeavor. Efforts to promote sustainable agriculture, implement stricter regulations on logging and mining, and invest in sustainable economic alternatives for local communities are crucial steps toward securing the Amazon's future. .

It is not too late to turn the tide, to protect this precious heritage for generations to come. We must rise to the challenge, embracing a sense of collective responsibility and acting decisively to safeguard the Amazon's future and ensure the well-being of our planet. .

The Amazon rainforest stands as a powerful reminder of the interconnectedness of life on Earth, a testament to the delicate balance that sustains our world. By embracing conservation as a core principle, we can ensure that this magnificent ecosystem continues to thrive, a source of wonder, resilience, and hope for generations to come. .

Chapter 2: The Canopy: Life Among the Branches

A World of Light and Air: Exploring the Canopy's Unique Environment

A World of Light and Air: Exploring the Canopy's Unique Environment.

The Amazon, a sprawling tapestry of emerald green, is often perceived as a dense, humid jungle, a place of perpetual shadow where life crawls and slithers on the forest floor. Yet, above the familiar symphony of the undergrowth, a vibrant world unfolds, a realm of filtered sunlight and gentle breezes, a haven for countless creatures adapted to a life high above the ground. This is the canopy, an intricate network of branches and leaves, a world of its own, teeming with life and brimming with unique ecological challenges.

Sunlight, the lifeblood of any ecosystem, filters through the canopy's leafy embrace, casting dappled patterns on the forest floor below. This mosaic of light and shade creates a dynamic environment where life thrives in its own unique niche. Some plants, like the towering Kapok tree, reach for the sun with their massive crowns, capturing the maximum amount of light. Others, like the delicate epiphytes, cling to branches, thriving in the filtered light that bathes the canopy. These specialized adaptations highlight the crucial role light plays in shaping the canopy's flora. .

But light is not the only factor that defines this aerial realm. The canopy is also a realm of air, a world of constant movement and change. The gentle breezes that rustle through the leaves carry with them the whispers of life, the scent of blossoming flowers, and the subtle changes in temperature and humidity. These aerial currents are crucial for seed dispersal, pollination, and even the regulation of the rainforest's climate. The canopy's intricate network of branches and leaves acts as a giant filter, mitigating the harsh tropical sun and regulating the flow of air, creating a unique microclimate within its embrace.

Within this complex tapestry of light and air, a fascinating array of creatures find their home. The canopy's abundance of fruits, flowers, and insects attracts a diverse range of birds, from the brilliant blue morpho butterfly to the majestic harpy eagle. Monkeys, with their nimble agility, navigate the branches with ease, their acrobatic feats a testament to their adaptation to this aerial world. Even mammals like sloths, whose slow, deliberate movements might seem ill-suited for such a dynamic environment, have found their niche in the canopy, clinging to branches and blending seamlessly into their leafy surroundings.

The canopy's unique environment has given rise to remarkable adaptations in its inhabitants. The spider monkeys, with their prehensile tails, use them as an extra limb, navigating the intricate branches with unmatched grace. The toucans, with their large, colorful beaks, feed on fruits that other birds cannot reach, playing a crucial role in seed dispersal. Even the humble ant, with its intricate social structure, creates complex, interconnected colonies that weave through the canopy, transporting food, building nests, and playing a vital role in the intricate web of life.

The canopy's unique environment is not only a haven for life but also a testament to the resilience of nature. The dense foliage acts as a natural barrier, protecting the fragile ecosystem from the destructive forces of human activities. Yet, despite its resilience, the canopy is under increasing threat. Deforestation, driven by human activities, is rapidly shrinking the rainforest's canopy, endangering the unique ecosystem and its inhabitants.

Exploring the canopy is not just a journey through a world of wonder but a journey of discovery. It reveals the intricate web of life that exists beyond the familiar ground, highlighting the delicate balance that sustains this vital ecosystem. The canopy's unique environment, with its interplay of light, air, and life, reminds us of the interconnectedness of all living things and the importance of protecting this precious resource for future generations. .

The Canopy's Inhabitants: Adaptations for Survival in the Heights

The Canopy's Inhabitants: Adaptations for Survival in the Heights.

The Amazon rainforest is a symphony of life, a tapestry woven with an intricate web of dependencies. Beneath the emerald expanse of the canopy, a hidden world unfolds. It is a realm of filtered light, swaying branches, and a constant struggle for existence. Here, amidst the rustling leaves and vibrant blooms, dwells a unique assemblage of creatures - the canopy's inhabitants - who have evolved extraordinary adaptations to thrive in this challenging environment. .

The canopy presents a unique set of ecological pressures, shaping the very essence of life within its embrace. Limited access to resources like water and soil necessitates innovative strategies for survival. The ever-present threat of predators and competitors demands agility, stealth, and specialized defenses. This aerial world, a realm of constant movement and precarious perches, necessitates adaptations that defy gravity and challenge the very definition of what it means to live.

The canopy's denizens are a testament to the power of natural selection, showcasing an astonishing array of evolutionary solutions. From the vibrantly colored and agile monkeys that swing effortlessly through the branches to the camouflaged insects that blend seamlessly with the foliage, each creature carries the mark of its adaptations. These adaptations, intricately interwoven with the physical and biological landscape of the canopy, are the key to understanding the complex dynamics of life in this remarkable ecosystem.

The Challenges of Living High:

The canopy, a world of dappled sunlight and swaying branches, is far from a comfortable haven. The constant struggle for survival presents a formidable set of challenges, demanding innovative solutions and remarkable adaptations.

Water Scarcity:

While the Amazon is synonymous with abundant rainfall, the canopy's inhabitants face a constant challenge – water scarcity. Rainwater, their primary source, is fleeting, quickly draining through the dense foliage. The creatures of the canopy must contend with this intermittent water supply, developing ingenious strategies to conserve and access this precious resource.

Limited Resources:

The canopy, a world of limited space and resources, necessitates a unique strategy for acquiring sustenance. Unlike the forest floor, where nutrient-rich soil supports a diverse array of plant life, the canopy relies on aerial resources - fruits, flowers, and insects. This presents a challenge for herbivores and carnivores alike, demanding adaptations for accessing these dispersed and often seasonal food sources.

Precarious Perches:

Life in the canopy is a delicate dance with gravity. Swaying branches and unsteady perches demand exceptional agility and balance. Creatures must adapt to navigating this dynamic environment, evolving specialized features to ensure their safety and mobility.

Predation and Competition:

The canopy, teeming with life, is also a crucible of predation and competition. From agile snakes to predatory birds, a variety of threats lurk in the branches, forcing its inhabitants to develop strategies for evasion, defense, and survival. The pressure to outcompete rivals for resources, territory, and mates adds another layer of complexity to the challenges of living in this dynamic environment.

The Canopy's Adaptations:

In response to these unique challenges, the canopy's inhabitants have evolved a remarkable array of adaptations. These adaptations, both physical and behavioral, showcase the extraordinary power of natural selection and the intricate interplay between life forms and their environment.

Physical Adaptations:

The canopy's inhabitants have evolved a range of physical adaptations that enable them to thrive in their unique environment. These adaptations, honed over millennia, reflect the delicate balance between form and function, reflecting the demands of their aerial existence.

Locomotion and Agility:

The canopy, a realm of swaying branches and precarious perches, demands exceptional agility and balance. The creatures of the canopy have evolved remarkable adaptations to navigate this dynamic environment, ranging from prehensile tails to specialized limbs and unique gaits.

Prehensile Tails: Many primates, like the spider monkey and the howler monkey, possess prehensile tails, which act as a fifth limb, providing an extra point of contact and allowing for greater maneuverability through the branches.

Specialized Limbs: Some canopy dwellers, like sloths and opossums, have evolved long, curved claws that enable them to grip tightly to branches, ensuring stability even as the tree sways in the wind.

Unique Gaits: The slow loris, with its remarkably slow and deliberate movements, and the gibbon, with its distinctive brachiation (arm swinging) gait, showcase the diversity of movement adaptations found in the canopy.

Camouflage and Defense:

The canopy is a dangerous place, teeming with predators and competitors. The creatures that call it home have evolved a variety of adaptations to avoid detection and defend themselves. .

Camouflage: Many insects, like stick insects and leaf insects, have evolved remarkable camouflage, blending seamlessly with the foliage, making them virtually invisible to predators.

Defense Mechanisms: Some animals, like the poison dart frog, possess toxic skin secretions, making them unpalatable to potential predators. Others, like the armadillo, have evolved protective shells for defense.

Warning Coloration: The vibrant colors of some animals, like the scarlet macaw and the poison dart frog, serve as a warning signal to potential predators, indicating that they are toxic or unpalatable.

Dietary Adaptations:

The canopy, a world of limited and often dispersed resources, necessitates specialized adaptations for acquiring sustenance. The creatures of the canopy have evolved unique dietary habits and digestive systems to exploit the available food sources.

Fruit-eating: Many canopy dwellers, like monkeys and toucans, are specialized fruit eaters, with adaptations like strong beaks and digestive systems that can break down tough fruits.

Insectivory: A wide array of birds and mammals, like woodpeckers and bats, are insectivores, equipped with specialized beaks or tongues to access insects within tree trunks and crevices.

Nectarivores: Some birds, like hummingbirds, are nectarivores, with long, slender beaks adapted to reach deep into flowers to access nectar.

Water Conservation:

The canopy, a world of fleeting rainfall, presents a unique challenge in terms of water conservation. The creatures of the canopy have evolved a variety of mechanisms to conserve and access this precious resource.

Metabolic Efficiency: Some animals, like sloths, have evolved a slow metabolism, reducing their need for water.

Dew Collection: Some insects, like certain species of beetles, have evolved specialized structures on their bodies to collect dew from the leaves.

Water Storage: Some animals, like the spiny-tailed lizard, store water in their tails, allowing them to survive periods of drought.

Behavioral Adaptations:

Beyond physical adaptations, the canopy's inhabitants have also evolved a range of behavioral adaptations, strategies, and social structures that enhance their survival in this challenging environment.

Social Structures:

Many canopy dwellers live in complex social groups, exhibiting intricate social hierarchies and communication systems. These social structures enhance survival by providing protection from predators, facilitating food acquisition, and enabling the efficient rearing of offspring.

Monkeys: Many monkey species live in troops with complex social hierarchies, allowing for coordinated defense against predators and efficient food gathering.

Birds: Some bird species, like parrots, live in flocks, providing greater vigilance against predators and allowing for more efficient food finding.

Communication:

Communication is essential for survival in the canopy, facilitating coordination, warning signals, and social interaction. The creatures of the canopy have evolved a wide range of communication signals, from vocalizations to visual displays.

Vocalizations: Many birds and mammals use a variety of calls and songs to communicate with each other, warning of threats, attracting mates, or maintaining social cohesion.

Visual Displays: Some animals, like the peacock, use elaborate visual displays to attract mates or intimidate rivals.

Territoriality:

Many canopy dwellers are highly territorial, defending their territories from rivals to secure access to resources and mates. Territoriality can be enforced through displays, vocalizations, and physical confrontations.

Birds: Many bird species defend territories, particularly during breeding season, to ensure access to nesting sites and food resources.

Mammals: Some mammal species, like jaguars, establish territories to protect their food sources and potential mates.

Nesting and Reproduction:

The canopy presents unique challenges for reproduction and offspring rearing. The creatures of the canopy have evolved various adaptations to ensure the survival of their young.

Nests: Many birds and mammals build nests in trees, providing a safe and sheltered environment for their young.

Parental Care: Many canopy dwellers exhibit extended parental care, providing protection, food, and social learning to their young.

The Canopy: A Tapestry of Life.

The canopy, a world of filtered light, swaying branches, and constant struggle, is a testament to the power of natural selection and the intricate interplay between life forms and their environment. The creatures that call this aerial world home have evolved a remarkable array of adaptations - physical and behavioral - enabling them to navigate the challenges of their unique habitat.

From the agile monkeys that swing effortlessly through the branches to the camouflaged insects that blend seamlessly with the foliage, each creature carries the mark of its adaptations. These adaptations, intricately woven with the physical and biological landscape of the canopy, are the key to understanding the complex dynamics of life in this remarkable ecosystem. .

The canopy, a vibrant and dynamic ecosystem, is a testament to the diversity and resilience of life on Earth. It is a reminder that even in the most challenging environments, life finds a way, evolving extraordinary solutions to conquer adversity and thrive in a world of constant change.

The Canopy's Role in the Ecosystem: A Vital Bridge for Life

The Amazon rainforest, a symphony of life pulsating across millions of acres, is often painted as a realm of lush green undergrowth, teeming with exotic creatures. While this image holds a kernel of truth, it only captures a fraction of the Amazon's true complexity. The heart of this ecosystem, the very beating pulse of its biodiversity, lies far above the shadowed forest floor - within the intricate tapestry of the canopy. This arboreal realm, a world of interlacing branches, vibrant foliage, and cascading vines, plays a pivotal role in the intricate dance of life that defines the Amazon. .

Imagine a city, not of concrete and steel, but of living, breathing trees, their crowns forming a dense, interconnected network that stretches across the horizon. This arboreal metropolis, the Amazonian canopy, is a microcosm of the entire ecosystem, supporting a staggering array of life forms, each playing a unique role in the grand scheme of survival. .

The canopy's influence begins with the very air we breathe. Its leaves, a vast photosynthetic factory, absorb the sun's energy and release oxygen, enriching the atmosphere not just for the rainforest, but for the planet itself. This constant oxygen production fuels the entire ecosystem, breathing life into the countless organisms that call the Amazon home. .

But the canopy's influence extends far beyond oxygen production. The very architecture of this arboreal realm, a complex network of branches and leaves, creates a unique microclimate, a haven for life. The canopy acts as a massive umbrella, filtering the intense sunlight, moderating the temperature, and regulating humidity. This controlled environment fosters a diverse range of habitats, each with its own unique set of flora and fauna. .

The canopy's structure also influences the flow of water within the rainforest. Its intricate network of leaves intercepts rainfall, slowing its descent and preventing rapid runoff. This process allows water to infiltrate the soil, replenishing the underground reservoirs that sustain the forest. Additionally, the canopy's intricate network of branches and leaves provides a surface for water to condense and form dew, a vital source of moisture for many canopy dwellers. .

The very act of breathing within the canopy is a testament to its crucial role. As leaves transpire, releasing water vapor into the air, they create a microclimate of high humidity, essential for the survival of numerous species. This humidity, combined with the constant temperature regulation, creates a stable environment that supports the growth of unique epiphytes, plants that grow on other plants, further enriching the canopy's biodiversity.

Imagine the canopy as a giant, intricate web of life, each thread connected to another, each strand a vital component of the ecosystem's intricate dance. This web is not merely a physical structure, but a complex network of interactions, a symphony of relationships between flora and fauna. .

Within this web, the canopy's intricate structure supports an incredible diversity of life, each organism playing a specific role in the ecosystem's delicate balance. Trees, the very foundation of the canopy, provide a physical platform for countless species. Their bark, branches, and leaves become homes, nesting sites, and feeding grounds for a myriad of organisms, from tiny insects to massive mammals. .

The leaves themselves, a cornucopia of colors and textures, serve as a crucial source of food and shelter. They are a canvas for an intricate tapestry of life, supporting countless insects, spiders, and other invertebrates. These creatures, in turn, become prey for a range of larger animals, creating a complex food web that weaves through the canopy's intricate architecture.

From the tiny, brilliantly colored butterflies that flit between leaves to the elusive monkeys that swing through the branches, the canopy is a tapestry of vibrant life. Each species, from the tiniest insect to the most majestic bird, plays a role in the canopy's intricate ecosystem. .

The canopy, a bustling metropolis of life, hums with activity, a symphony of sounds and movements, a testament to its dynamic nature. From the rustling of leaves in the breeze to the calls of monkeys echoing through the branches, the canopy is a symphony of life, each sound a unique note in the rainforest's grand melody. .

Beyond its sheer biodiversity, the canopy's influence extends to the very fabric of the Amazonian ecosystem. It acts as a conduit for nutrients, a crucial bridge between the forest floor and the upper atmosphere. The decomposition of fallen leaves and branches on the forest floor provides vital nutrients, which are then carried upwards by insects, birds, and other animals, enriching the canopy's soil and promoting the growth of new life. .

This constant cycling of nutrients, a vital process facilitated by the canopy, ensures the Amazon's long-term health and resilience. It is a constant dance of life and death, a cycle of growth and decay that ensures the forest's ongoing vitality. .

The canopy, a crucial component of the Amazon's intricate ecosystem, acts as a bridge, connecting the forest floor to the upper atmosphere, and in doing so, linking the entire ecosystem into a delicate network of interconnected life. .

However, this intricate web of life is increasingly threatened by human activities. Deforestation, a relentless assault on the Amazon's pristine forests, is leading to the loss of countless tree species, severing the vital link between the canopy and the forest floor. The consequences of this loss are far-reaching, disrupting the delicate balance of the Amazonian ecosystem and endangering the lives of countless species. .

As the Amazon's canopy shrinks, so does its ability to regulate the climate, moderate rainfall, and provide habitat for its unique biodiversity. The loss of this vital bridge threatens not only the Amazon's future, but the well-being of the planet as a whole.

The Amazon's canopy, a vibrant tapestry of life, stands as a testament to the power and complexity of nature. Its intricate structure and diverse inhabitants are a constant reminder of the delicate balance that defines our planet's ecosystems. Understanding the canopy's vital role is not just a matter of scientific curiosity; it is a crucial step in ensuring the Amazon's continued vitality and the future of our planet.

Exploring the Canopy: Methods of Research and Exploration

Exploring the Canopy: Methods of Research and Exploration.

The Amazon rainforest, a verdant tapestry woven across the South American landscape, holds within its depths a realm of unparalleled biodiversity. Its canopy, a living cathedral of emerald and sunlight, shelters a dazzling array of life, a world largely unexplored and veiled in mystery. Unveiling this hidden kingdom requires a blend of ingenuity, persistence, and a deep understanding of the delicate ecosystem that thrives within its branches. This exploration demands a unique set of research methods, pushing the boundaries of scientific inquiry and forging new paths into the heart of this vibrant, aerial world.

Traditionally, the dense, tangled undergrowth of the Amazon has presented a significant barrier to accessing its arboreal realm. Early expeditions relied primarily on ground-based observations, capturing glimpses of the canopy's inhabitants through binoculars or by climbing the tallest trees within reach. Yet, this offered a limited perspective, only scratching the surface of the intricate web of life flourishing above. The advent of new technologies, however, has revolutionized our ability to explore the Amazon's canopy, enabling us to delve deeper into its secrets. .

One of the most significant advancements has been the development of canopy walkways, or "sky bridges," which provide a safe and stable platform for scientists to traverse the upper reaches of the forest. These elevated structures, ranging from simple rope bridges to sophisticated, multi-tiered walkways, offer unprecedented access to a previously inaccessible environment. Researchers can now spend hours observing, collecting data, and conducting experiments in the heart of the canopy, fostering a deeper understanding of the intricate relationships between its diverse inhabitants.

Beyond walkways, climbing techniques have also played a crucial role in exploring the canopy. Experienced climbers, equipped with specialized gear and a profound understanding of arboreal safety, can ascend to the highest branches, allowing for close-up observations and meticulous data collection. This technique, while demanding physical prowess and meticulous attention to detail, provides unparalleled access to the canopy's most secluded areas, offering glimpses into the lives of rare and elusive species.

However, even the most adept climbers face limitations. The sheer scale and complexity of the Amazonian canopy demand a more comprehensive approach. This is where remote sensing technologies come into play. LiDAR (Light Detection and Ranging), for instance, utilizes laser pulses to create detailed three-dimensional models of the canopy, revealing its intricate structure and identifying potential areas of interest for further exploration. Aerial photography and satellite imagery provide a broader perspective, mapping the extent and distribution of different canopy types and highlighting areas with high biodiversity.

These technologies offer a powerful toolset for studying the canopy's structural complexity, but exploring the intricate web of life that thrives within requires a more direct approach. Camera traps, strategically placed throughout the canopy, capture images and videos of elusive animals, revealing patterns of behavior and shedding light on their ecological roles. Acoustic monitoring utilizes specialized microphones to record the sounds of the forest, revealing the presence of vocal animals and offering insights into their communication patterns.

Further understanding the dynamics of this complex ecosystem requires biological sampling, a delicate balancing act between scientific curiosity and ecological responsibility. Leaf litter analysis, for example, reveals information about the diet and habitat preferences of canopy inhabitants. DNA sequencing of soil and water samples can identify the presence of specific species, even those elusive or difficult to observe directly. Isotopic analysis of animal tissues provides information about their feeding habits and movement patterns within the canopy.

The study of the Amazon's canopy is not limited to physical exploration. Ethnobotanical research plays a crucial role in understanding the complex relationship between humans and the forest. By engaging with indigenous communities who have inhabited the Amazon for centuries, scientists can tap into a wealth of traditional knowledge regarding the use of canopy resources, the medicinal properties of its plants, and the ecological significance of its various species. This knowledge, passed down through generations, provides a valuable perspective on the intricate web of life that binds the canopy and its human inhabitants.

The research methods employed in exploring the Amazon's canopy are constantly evolving, driven by technological advancements, a deep understanding of the forest's delicate balance, and a growing appreciation for the immense biodiversity that thrives within its branches. Each new discovery, each innovative approach, unravels another layer of the complex tapestry of life, bringing us closer to comprehending the secrets hidden within this verdant cathedral of the sky. The journey into the Amazon's canopy is a testament to the enduring human desire to explore, discover, and protect the world's most extraordinary natural wonders. .

Chapter 3: The Understory: A World of Shadows and Shade

The Understory's Environment: A Realm of Constant Change

The Understory's Environment: A Realm of Constant Change.

The Amazon rainforest, a vast tapestry of life woven across the South American continent, is a realm of dramatic contrasts, nowhere more so than in its understory. This shadowy, verdant world, perpetually bathed in dappled sunlight, presents a stark juxtaposition to the sun-drenched canopy above. Yet, beneath the seemingly tranquil façade, a tempestuous drama unfolds, a continuous battle for survival where every inch of space and every fleeting ray of light is contested. Here, the environment is not a static backdrop but a dynamic force, shaping and reshaping the lives of countless organisms.

The understory, a world of perpetual twilight, is defined by its limited access to sunlight. While the canopy above basks in the full glory of the sun, the understory exists in a world of perpetual penumbra. This constant state of low-light conditions has shaped the evolution of the understory's flora, favoring plants with broad leaves, adapted for efficient photosynthesis in low light. The understory's flora is also characterized by a variety of climbing plants, clinging to larger trees in their quest for access to the precious sunlight above. These adaptations, driven by the understory's limited light, have created a unique and intricate ecosystem teeming with life.

But the understory's struggle for light is just one facet of its ceaseless change. The very air it breathes is a constant source of fluctuation. Its humidity, perpetually high, fluctuates with the rhythm of the seasons, creating a dynamic environment where moisture is both abundant and precious. The air is thick with the scents of decaying vegetation, the aroma of the forest floor, a testament to the ongoing cycle of life and death. These scents are more than just olfactory sensations; they are powerful chemical signals, broadcasting messages of decay, growth, and predation, weaving the understory into a complex network of communication.

The understory's soil, a rich tapestry of decaying organic matter, is a constant source of nutrients and a crucible of change. It is a haven for a staggering diversity of fungi and bacteria, engaged in an intricate dance of decomposition and nutrient cycling. Their ceaseless activity fuels the understory's growth, providing the foundation for the forest's unparalleled biodiversity. This rich soil is also a vital component of the understory's dynamic environment, its composition shifting with the seasons, impacted by rainfall, and influenced by the ceaseless movement of organisms.

But the understory is not just a place of constant change; it is also a place of constant movement. The forest floor, covered in a mosaic of fallen leaves, decaying logs, and tangled roots, is a bustling arena of activity. Animals, both large and small, navigate this labyrinthine landscape, their movements shaping the understory's environment. The rustling of leaves, the thudding of hooves, the screech of a monkey in the canopy above, these are all sounds that punctuate the understory's quiet symphony of change.

The understory is also a place of constant competition. Its resources, limited by the perpetual shade and the intricate web of interdependencies, are fiercely contested. Plants compete for sunlight, nutrients, and space. Animals compete for food, shelter, and mates. This constant struggle for survival, a defining characteristic of the understory, shapes the evolution of its inhabitants, pushing them to adapt, innovate, and thrive.

This competition for resources is amplified by the ever-present threat of predation. The understory, with its dense foliage and limited visibility, is a hunting ground for a diverse array of predators. From the stealthy jaguar to the venomous snake, from the agile spider to the patient frog, predators constantly stalk their prey, adding another layer of complexity to the understory's dynamic ecosystem.

But even amidst the struggle for survival, there is beauty and wonder. The understory, with its muted light and the intricate tapestry of life, is a place of stunning visual contrasts. The vibrant hues of flowers, the patterns of leaf veins, the shimmering scales of a lizard sunning itself on a fallen log, all combine to create a scene of unparalleled beauty. The understory is a place where life bursts forth in a kaleidoscope of forms, a testament to the resilience and adaptability of nature.

This constant change, a defining characteristic of the understory, is not a sign of instability but rather a hallmark of its vitality. It is the driving force behind its biodiversity, its resilience, and its enduring beauty. It is a testament to the interconnectedness of life, a reminder that change, however dramatic, is not just a force of destruction but also a force of creation.

The understory, a world of shadows and shade, is a constant reminder that even in the most seemingly tranquil environments, life is a dynamic, ever-evolving force. The battle for survival, the constant flux of resources, the ceaseless dance of predator and prey, all contribute to the understory's vibrant tapestry of life. This realm of constant change, a place of both struggle and beauty, holds within its depths a profound testament to the extraordinary power of nature's resilience. .

Adaptation and Survival in the Understory: A Challenge for Life

The Amazon rainforest, a colossal tapestry of life woven over millions of years, presents an array of challenges and opportunities for its inhabitants. While the canopy, bathed in the sun's golden rays, is often celebrated for its abundance and grandeur, the understory, a realm of perpetual twilight, harbors an equally remarkable, albeit often overlooked, diversity. This shadowy world, shrouded in a perpetual dappled light, demands a unique set of adaptations for survival. .

Within the understory, the struggle for resources is fierce. Sunlight, the lifeblood of the forest, is scarce, filtered through a dense canopy of leaves. Plants, forced to compete for limited light, have evolved ingenious strategies. Some, like the towering kapok trees, reach their branches skyward, their massive trunks serving as pillars of support. Others, like the strangler figs, are epiphytes, growing on the branches of larger trees, their roots snaking down to the forest floor, eventually enveloping their host. Still others, like the countless species of ferns, have adapted to thrive in the dim, humid conditions, their fronds unfurling in a dance of green. .

The fight for light is not limited to plants. Animals too, have adapted to this challenging environment. Many, like the elusive jaguar, are primarily nocturnal, emerging from the shadows under the cloak of darkness to hunt. Others, like the brightly colored toucans, have evolved long, sharp beaks to reach fruits high in the canopy, often inaccessible to their terrestrial competitors. The intricate web of life in the understory is further complicated by the constant struggle for food. The forest floor, covered in a layer of decaying leaves and branches, is teeming with invertebrates, providing a vital food source for a variety of creatures. .

The understory, however, is not just a battleground for survival. It is also a haven for life. Its dense vegetation provides shelter for a myriad of creatures, from the tiny insects that flit among the leaves to the large mammals that roam the forest floor. The abundance of decaying matter provides sustenance for fungi and bacteria, playing a crucial role in the forest's ecosystem. The intricate network of roots, intertwined like a subterranean labyrinth, creates a complex ecosystem, supporting a vast array of organisms. .

The understory, with its unique set of challenges and opportunities, is a testament to the incredible resilience of life. It is a world of adaptation and survival, a constant struggle for resources, a delicate balance between competition and cooperation. It is a microcosm of the Amazon's vast biodiversity, a hidden world teeming with life, where every organism plays a vital role in the delicate dance of the rainforest ecosystem. .

Key Players in the Understory: Animals and Plants of the Shade

The Amazon rainforest, with its colossal canopy reaching towards the heavens, casts a world of dappled light and perpetual twilight below. This is the understory, a realm of hushed whispers and subtle beauty, where life flourishes in the absence of direct sunlight. Here, a unique cast of characters has evolved, mastering the art of survival in this dimly lit, yet fertile world. The understory is not merely a supporting act to the grand spectacle above; it is a vibrant ecosystem unto itself, pulsating with the silent drama of life in the shade.

The understory's denizens are masters of adaptation. Plants, deprived of the full sun's embrace, have developed intricate strategies to thrive in the meager light that penetrates the dense canopy. Some have elongated stems that reach towards the upper tiers, stretching for precious rays of sunlight. Others have developed large, broad leaves that efficiently capture what little light is available. The intricate patterns of leaves and the mosaic of light and shadow create an ever-shifting canvas of beauty, a testament to the ingenuity of nature's architects.

Among the understory's botanical denizens are the fascinating epiphytes, plants that grow on other plants, drawing sustenance not from the soil but from the air and the debris that collects on their host. Orchids, bromeliads, and ferns, often adorned with vibrant hues and intricate structures, drape themselves upon the trunks and branches of the giants above. Their aerial roots, clinging tenaciously to their perches, absorb moisture and nutrients from the humid air, creating miniature gardens that contribute to the understory's vibrant tapestry.

The understory is also a haven for a remarkable array of animals, each exquisitely adapted to the unique challenges of this dimly lit world. The forest floor is a carpet of decaying leaves and fallen branches, providing a rich source of nutrients and a haven for invertebrates, amphibians, and reptiles. Lizards, snakes, and frogs, often brilliantly colored, dart through the undergrowth, their keen senses honed to detect prey and avoid predators. The air hums with the chorus of insects, their buzzing and chirping a symphony of life in the shade.

Many understory inhabitants are masters of camouflage, their coloration and patterns blending seamlessly with the dappled light and mottled leaves. This is especially true for predators like jaguars and pumas, who stalk their prey through the undergrowth, their tawny coats mirroring the shadows and sunlight filtering through the canopy. The understory's inhabitants have evolved a complex web of predator-prey relationships, where every creature, from the smallest insect to the largest mammal, plays a crucial role in the delicate balance of life.

A remarkable diversity of primates, too, call the understory home. Monkeys, like the agile spider monkey and the powerful howler monkey, swing through the upper branches, their calls echoing through the rainforest canopy. But some primates, like the nocturnal night monkey, have adapted to the understory's darkness, their senses attuned to the symphony of sounds that fill the night. These primates, along with the other understory dwellers, are crucial players in the forest's complex web of life, dispersing seeds, pollinating flowers, and contributing to the intricate tapestry of this unique ecosystem.

The understory is a realm of whispers, of rustling leaves and the soft chirping of insects, where life unfolds at a slower pace. It is a world of hidden wonders, of delicate beauty and intricate relationships. The animals and plants that inhabit this shadowy realm have developed extraordinary adaptations, allowing them to thrive in the absence of direct sunlight. The understory is not merely a shadowy backdrop to the rainforest's grand spectacle, but a vibrant ecosystem in its own right, a testament to the boundless creativity of nature.

The understory's inhabitants are not mere spectators in the drama of life, but active participants, shaping the rainforest's destiny. Their delicate web of interactions, from the intricate dance of pollination to the complex cycles of predation, ensures the rainforest's health and stability. The understory is a microcosm of life, a testament to the beauty and resilience of nature, and a reminder that even in the darkest corners of the rainforest, life finds a way to flourish.

The Understory's Floral Diversity.

The understory is a kaleidoscope of botanical diversity, where plants have evolved a myriad of adaptations to thrive in the meager light that filters through the rainforest canopy. One of the most striking features of the understory flora is the abundance of epiphytes, plants that grow on other plants, drawing sustenance not from the soil but from the air and the debris that collects on their host.

The orchids, with their intricate beauty and captivating fragrances, are among the most well-known epiphytes. Their roots, often clinging to the bark of trees, absorb moisture and nutrients from the air and the organic matter that accumulates around them. The delicate petals of orchids come in a stunning array of colors and patterns, attracting pollinators with their seductive allure. The bromeliads, too, are masters of the aerial life, their spiky leaves forming a rosette that collects rainwater and debris, providing a miniature ecosystem for frogs, insects, and other small creatures.

The understory is also home to a rich variety of ferns, their delicate fronds unfurling like emerald lace in the dim light. Many ferns, like the staghorn fern, are epiphytes, attaching themselves to the bark of trees with their aerial roots. Others, like the tree fern, are towering giants, their massive trunks providing a vertical habitat for a variety of other plants and animals. .

The Understory's Faunal Symphony.

The understory is alive with a symphony of sounds, a vibrant chorus of insects, amphibians, and reptiles. The forest floor, a carpet of decaying leaves and fallen branches, provides a haven for a diverse array of creatures, each playing a crucial role in the ecosystem's delicate balance.

The air hums with the buzzing of insects, a constant soundtrack to life in the understory. Ants, beetles, and butterflies flit through the undergrowth, their movements a blur of activity. These tiny creatures are essential for the forest's health, decomposing organic matter, dispersing seeds, and providing food for larger animals.

Among the understory's most fascinating denizens are the amphibians, their moist skin and delicate bodies perfectly adapted to the humid environment. Frogs, with their vibrant colors and croaking calls, are a familiar sight in the understory. Some, like the poison dart frog, are brightly colored, a warning to potential predators of their toxic nature. Others, like the glass frog, are translucent, their bodies blending seamlessly with the leaves on which they perch.

Reptiles, too, play a crucial role in the understory's web of life. Lizards, with their quick movements and keen eyes, dart through the undergrowth, catching insects and small invertebrates. Snakes, their bodies sleek and sinuous, slither through the fallen leaves, their forked tongues tasting the air for prey. The understory is a fascinating microcosm of life, a vibrant tapestry woven from the interactions of plants and animals, each playing its unique part in the ecosystem's delicate balance.

The Understory's Predators.

The understory is not without its predators, creatures who have honed their skills to hunt in the dim light and navigate the dense undergrowth. Among these are the jaguars, the apex predator of the Amazon rainforest, their tawny coats blending seamlessly with the dappled shadows of the understory. Their keen senses, powerful muscles, and stealthy movements make them formidable hunters, capable of taking down prey much larger than themselves. .

Pumas, too, are masters of the understory, their agile bodies and sharp claws allowing them to navigate the dense undergrowth with ease. These solitary hunters, known for their stealth and cunning, prey on a variety of animals, from deer and peccaries to smaller mammals like monkeys and rodents.

The understory is a dangerous place, but it is also a place of incredible beauty and resilience. The animals and plants that inhabit this shadowy realm have developed extraordinary adaptations, allowing them to thrive in the absence of direct sunlight. Their delicate web of interactions, from the intricate dance of pollination to the complex cycles of predation, ensures the rainforest's health and stability. .

The Understory's Nightlife.

As the sun dips below the horizon, the understory transforms, emerging from the shadows as a world of nocturnal activity. The forest floor, now bathed in the soft glow of the moon, comes alive with the sounds of insects and the rustle of nocturnal animals.

The night monkey, a small, nocturnal primate, is a fascinating inhabitant of the understory. Its large, expressive eyes, adapted to seeing in the dim light, allow it to navigate the forest floor with ease. These primates, unlike their diurnal counterparts, are active at night, foraging for fruits and insects in the understory's dimly lit realm.

The understory is also home to a variety of nocturnal birds, like the potoo and the nightjar, whose cryptic coloration blends seamlessly with the surrounding foliage. These birds, often active during the hours of darkness, prey on insects and other small creatures, their sharp eyes and sensitive ears allowing them to detect prey in the dimly lit understory. .

The understory's nightlife is a fascinating glimpse into the hidden world of the rainforest, a realm where life unfolds at a different pace, governed by the rhythms of the moon and stars. The nocturnal denizens of the understory, with their specialized adaptations and unique ways of life, remind us of the incredible diversity and resilience of nature.

The Understory's Importance.

The understory is not merely a shadowy backdrop to the rainforest's grand spectacle; it is a vibrant ecosystem in its own right, playing a crucial role in the health and stability of the Amazon rainforest. It is a reservoir of biodiversity, harboring a wealth of plant and animal life, each playing its unique part in the delicate balance of this complex ecosystem.

The understory's plants and animals are intimately interconnected, their interactions forming a complex web of life that sustains the rainforest. Plants provide food and shelter for animals, while animals pollinate plants, disperse seeds, and help to cycle nutrients through the ecosystem. The understory's diverse array of plants and animals are vital for the rainforest's ability to absorb carbon dioxide, regulate the climate, and provide habitat for a vast array of species.

The understory is also a crucial source of food and medicine for local communities. Many traditional medicines are derived from plants found in the understory, while its fruits, nuts, and other resources provide sustenance for indigenous peoples. The understory is a testament to the power of nature, a reminder of the interconnectedness of life and the importance of preserving biodiversity.

The Understory's Future.

The understory, like the rainforest itself, is facing unprecedented challenges. Deforestation, habitat loss, and climate change are threatening the delicate balance of this vital ecosystem. As the rainforest canopy is cleared for agriculture, logging, and mining, the understory is often destroyed, depriving countless plants and animals of their homes and resources.

Climate change is also taking a toll on the understory, with rising temperatures and altered rainfall patterns impacting the delicate balance of this complex ecosystem. The understory, with its unique adaptations and sensitive nature, is particularly vulnerable to these changes.

The future of the understory is uncertain, but it is clear that its preservation is crucial for the health and stability of the Amazon rainforest. Protecting this vibrant ecosystem is not just about preserving biodiversity; it is about safeguarding the future of the planet, its climate, and the countless lives that depend on its delicate balance.

A Call to Action.

The understory, with its hidden wonders and delicate beauty, is a reminder of the incredible diversity and resilience of nature. It is a testament to the power of adaptation, a world where life finds a way to flourish even in the most challenging of environments. But the understory, like the rainforest itself, is under threat, facing an array of challenges that threaten its very existence.

The future of the understory, and indeed the future of the Amazon rainforest, depends on our actions. We must work to protect this vital ecosystem, to reduce our impact on the environment, and to promote sustainable practices that ensure the health and resilience of the rainforest for generations to come. Let us celebrate the wonders of the understory, and let us work to ensure that its vibrant tapestry of life continues to flourish for years to come.

The Understory's Importance: A Nursery for Life

The Understory's Importance: A Nursery for Life.

The Amazon rainforest, a symphony of life, pulsates with vibrant green, a testament to the sheer power of nature's tapestry. But beneath the canopy, where sunlight struggles to penetrate, a different story unfolds – a world of shadows and shade, teeming with life. This is the understory, a realm of hidden wonders, a vital nursery for the rainforest's incredible biodiversity.

Imagine a world where dappled sunlight filters through a dense network of leaves and branches, creating a mosaic of light and dark. This is the understory, a realm of perpetual twilight, where life has adapted to thrive in a world of muted light and constant humidity. Here, the air is thick with the scent of damp earth and decaying leaves, a symphony of scents that speak of life and decay. This is the cradle of the Amazon's biodiversity, where delicate ferns unfurl their fronds, vibrant orchids bloom in unexpected places, and a menagerie of creatures, from colorful frogs to elusive jaguars, carve out their existence.

The understory is a haven for life. It provides a safe haven for countless species, offering protection from predators and harsh sunlight. Its moist environment, a haven of humidity, facilitates growth and provides a stable temperature, creating a perfect incubator for life. From the towering trees that form the canopy to the delicate seedlings that sprout at the forest floor, the understory is a vibrant ecosystem, a symphony of interconnectedness.

The understory is not merely a refuge for life; it is a crucial stage in the rainforest's life cycle. It is a nursery, a place where seedlings germinate, grow, and mature, nurtured by the rich organic matter of the forest floor. This vital layer of decomposing leaves and fallen branches provides essential nutrients, a vital lifeline for the rainforest's flora. As these seedlings mature, they rise through the canopy, eventually taking their place in the rainforest's grand symphony of life.

This symbiotic relationship, where the understory fosters the next generation, highlights the delicate balance of the rainforest. The understory plays a crucial role in the rainforest's resilience, ensuring its survival against environmental pressures. It serves as a buffer, protecting the rainforest from the harsh sun, absorbing excess rainfall, and mitigating the impact of drought. It is a testament to the intricate web of life that binds the rainforest together.

The understory is a vital reservoir of biodiversity. From the tiny insects that scurry through the leaf litter to the larger mammals that roam its hidden pathways, the understory teems with life. This diverse community plays a vital role in the rainforest's ecosystem, influencing nutrient cycling, pollination, and seed dispersal. The understory is a living laboratory, a microcosm of the rainforest's complexity and richness.

The understory is not just a place of shadows and shade; it is a dynamic world, constantly evolving and adapting to the changing environment. Its very existence is a testament to the rainforest's remarkable ability to regenerate and sustain life. This resilience, this capacity for renewal, is a testament to the power of nature, a force that cannot be underestimated.

The understory, a world hidden in plain sight, is a vital part of the Amazon's ecosystem. It is a nursery for life, a haven for biodiversity, and a testament to the rainforest's resilience. As we delve deeper into this hidden world, we gain a greater appreciation for the intricate web of life that sustains the Amazon, a symphony of nature that we must strive to protect. .

Chapter 4: The Forest Floor: Life in the Dark

The Forest Floor's Environment: A World of Decomposition and Rebirth

The Forest Floor's Environment: A World of Decomposition and Rebirth.

The forest floor, a world shrouded in perpetual twilight, is often overlooked in the grandeur of the towering canopy. Yet, this seemingly inert expanse is a teeming ecosystem, pulsating with life and a relentless cycle of decomposition and rebirth. Here, the sun's rays struggle to penetrate the dense foliage, painting the ground in a mosaic of light and shadow. The air, thick with the scent of decaying leaves and moist earth, is a symphony of rustling insects and the steady thump of falling fruits. This is the realm of the decomposers, the unseen architects of the forest's vitality.

The fallen leaves, branches, and logs, a constant rain of organic matter from the upper world, become the foundation of this hidden ecosystem. Their decay fuels a complex network of organisms, each playing a crucial role in breaking down the complex molecules into simpler forms. The process starts with the legions of detritivores, tiny creatures that feast on the dead matter. Earthworms, millipedes, and termites, working tirelessly beneath the surface, churn and break down the debris, creating a rich, nutrient-rich soil.

Fungi, with their intricate mycelial networks, play a crucial role in this decomposition. They secrete enzymes that break down lignin and cellulose, the tough building blocks of wood, making them accessible to other organisms. These intricate webs of fungal hyphae, like invisible veins, spread throughout the forest floor, connecting decaying matter to the roots of living trees. This symbiotic relationship provides the trees with essential nutrients, while the fungi receive sugars produced by the tree during photosynthesis.

Beyond the detritivores, a diverse array of invertebrates inhabit the forest floor. Beetles, ants, and spiders, each with their unique adaptations, contribute to the decomposition process. Some, like scarab beetles, bury dead animals, while others, like ants, transport dead leaves and insects to their nests, further breaking them down. This constant activity creates a rich, fertile substrate, vital for the growth of plants.

However, the forest floor is not just a graveyard. It is also a nursery, a cradle for new life. Seeds, dropped from the canopy above, find fertile ground in the rich, decaying litter. The moist, nutrient-rich environment provides the ideal conditions for germination and growth. The constant breakdown of organic matter releases vital nutrients, feeding the saplings that will eventually become the giants of the forest.

The forest floor also serves as a habitat for a myriad of creatures, large and small. Lizards, snakes, and frogs seek refuge amongst the fallen leaves and logs, blending into the dappled light. Small mammals, like rodents and opossums, scavenge for food and shelter amongst the undergrowth. These creatures, in turn, become prey for larger predators, creating a complex food web that helps regulate the ecosystem.

The forest floor's delicate balance is constantly threatened by human activities. Deforestation, logging, and agricultural expansion disrupt the natural cycles of decomposition and regeneration. The removal of trees disrupts the flow of nutrients and the microclimate of the forest floor, leading to soil erosion and loss of biodiversity.

The importance of the forest floor cannot be overstated. This seemingly humble layer is the heart of the Amazon's ecosystem, providing essential nutrients for the forest's growth, supporting an incredible diversity of life, and playing a vital role in regulating the global climate. It is a testament to the interconnectedness of nature, where the cycle of death and rebirth ensures the continued vitality of the rainforest. As we navigate the depths of the Amazon, we must remember that beneath our feet lies a world teeming with life, a world that deserves our respect and protection.

Life on the Forest Floor: Adaptations for Low-Light and Moisture

Life on the Forest Floor: Adaptations for Low-Light and Moisture.

The forest floor, a realm shrouded in perpetual twilight, is a world of contrasts. It is a haven of moisture, damp and humid, yet an arena of fierce competition for light, nutrients, and space. Life here thrives in a delicate balance, each organism exquisitely adapted to survive in this unique environment.

The dim, dappled light that filters through the dense canopy presents a formidable challenge. Plants must capture every photon, and animals must navigate and hunt in near-darkness. Plants on the forest floor have evolved an array of strategies to maximize their photosynthetic efficiency. Broad, flat leaves, like those of the towering Amazon water lily, maximize their surface area for light absorption. Other plants, such as the aptly named "shade-tolerant" species, possess a higher concentration of chlorophyll, enabling them to capture the sparse light more effectively. Some, like the herbaceous ground cover plants, even develop leaves with a mosaic pattern, allowing them to capture light from multiple angles.

Beyond light, the forest floor presents another challenge: moisture. Abundant rainfall and high humidity create a perpetually damp environment. Plants, like the epiphytic orchids, have evolved to thrive in this humid environment, drawing moisture from the air through their roots. Others, such as the terrestrial bromeliads, develop reservoirs within their leaf bases, collecting rainwater and creating mini-ecosystems within their foliage. These adaptations allow them to flourish in the perpetually damp conditions of the forest floor.

Yet, moisture also poses a challenge, increasing the risk of fungal diseases and decomposition. This constant threat has driven the evolution of protective mechanisms. The leaves of many plants are coated with a waxy layer, preventing excessive water absorption and fungal growth. Additionally, a unique array of fungal species has evolved to thrive in this damp environment, playing a crucial role in decomposition and nutrient recycling, further shaping the character of the forest floor ecosystem.

Animals, too, have adapted to the unique conditions of the forest floor. Arthropods, particularly insects, dominate this realm. Many have developed elongated limbs and antennae, allowing them to navigate through the dense undergrowth. Their ability to survive on a diverse range of decaying matter, from leaves to animal carcasses, makes them essential components of the forest floor ecosystem. .

Some animals, like the leaf-litter frog, have developed exceptional camouflage, blending seamlessly into the forest floor's mosaic of decaying leaves and fallen branches. Others, like the aptly named "floor-dwelling" spider, have woven intricate webs among the dense undergrowth, waiting patiently for their prey to stumble into their meticulously constructed traps.

The forest floor, however, is not just a realm of darkness and dampness. It is a vibrant tapestry of life, a bustling ecosystem teeming with activity. The decaying organic matter provides sustenance to a vast array of detritivores, from millipedes and earthworms to beetles and ants. These creatures, in turn, provide food for predators, creating intricate food webs that underpin the entire forest floor ecosystem.

The intricate interplay of light, moisture, and competition has shaped the unique adaptations of the forest floor's inhabitants. This environment, a symphony of survival and symbiosis, is a testament to the resilience and diversity of life in the Amazon jungle. Every organism, from the smallest insect to the largest tree, plays a vital role in the intricate web of life that defines this unique ecosystem. .

Decomposers and their Role: Breaking Down and Recycling Nutrients

The Amazon rainforest, a vibrant tapestry of life, pulsates with a ceaseless rhythm of creation and destruction. While the canopy, bathed in sunlight, boasts of dazzling colors and bustling activity, the forest floor, shrouded in perpetual twilight, plays a silent yet vital role in this grand ecosystem. It is here, in the realm of shadows, that a hidden army of decomposers tirelessly break down fallen leaves, decaying wood, and the remains of countless organisms, fueling the intricate web of life above. .

Decomposers, a diverse group of fungi, bacteria, and invertebrates, are the unsung heroes of the forest, performing an essential service that ensures the continuous flow of nutrients. They are the recycling machines of nature, converting complex organic matter into simpler forms that can be absorbed by plants, thus completing the cycle of life.

The forest floor, a carpet of decaying vegetation and fallen branches, is a treasure trove for decomposers. Imagine stepping onto this soft, spongy ground, the air heavy with the scent of damp earth and decaying leaves. Beneath your feet lies a rich tapestry of life, a bustling community of unseen organisms, each playing a crucial role in the grand scheme of nutrient cycling.

Fungi, the masters of decomposition, are the first to colonize dead organic matter. They secrete powerful enzymes that break down complex molecules like cellulose and lignin, the building blocks of plant tissues. These enzymes, like tiny molecular scissors, snip the long chains of organic molecules into smaller, more easily digestible pieces. The decomposition process is a dance between fungi and bacteria, each relying on the other to complete the task.

Bacteria, microscopic yet mighty, are the final stage in the breakdown of organic matter. They consume the simpler molecules released by fungi, converting them into even simpler forms that can be absorbed by plants. These nutrients, like nitrogen, phosphorus, and potassium, are then taken up by plants, forming the basis of new life.

The intricate dance of decomposers on the forest floor is a masterpiece of natural engineering. Imagine the sheer volume of organic matter that falls from the canopy each year - leaves, branches, fruits, and the carcasses of countless animals. If this organic matter were to accumulate indefinitely, the forest would be buried under a thick layer of debris, suffocating the life beneath.

But nature has a solution. Decomposers, fueled by the energy in organic matter, diligently break down this debris, returning vital nutrients to the soil, making them available for the next generation of plants. This continuous cycle, a delicate balance of life and death, ensures the continued productivity and resilience of the Amazon rainforest.

Consider, for instance, the role of termites in the decomposition of wood. These tiny insects, often overlooked, are vital in breaking down deadwood, returning nutrients to the soil and creating spaces for other decomposers to flourish. Their tunnels and galleries aerate the soil, improving drainage and water infiltration, making it more hospitable for plant roots.

Even the seemingly insignificant earthworms, constantly burrowing and churning the soil, contribute significantly to decomposition. They ingest organic matter, breaking it down further and mixing it with the soil, improving its structure and nutrient content. Their tunnels create pathways for air and water, facilitating the growth of plant roots.

The importance of decomposers goes beyond nutrient cycling. They also play a vital role in controlling disease outbreaks. By breaking down dead organic matter, they prevent the buildup of disease-causing pathogens, keeping the forest healthy. .

Their work is not without challenges. The forest floor, a world of darkness and humidity, presents unique obstacles. The lack of sunlight restricts the activity of many organisms, while the constant humidity can lead to the growth of harmful fungi and bacteria. But decomposers have adapted to these challenges, developing intricate mechanisms to survive and thrive in this unique environment.

One fascinating example of this adaptation is the symbiotic relationship between fungi and plants. Many fungi form mycorrhizal associations with tree roots, exchanging nutrients for sugars produced by the plant. This mutually beneficial relationship allows both partners to access resources that would otherwise be unavailable, ensuring their survival and growth.

The decomposition process is not a passive, isolated event. It is a dynamic interplay between organisms, each contributing to the breakdown of organic matter. Fungi and bacteria, along with invertebrates like termites and earthworms, work together in a complex web of interactions, ensuring the efficient and rapid breakdown of organic matter.

This intricate dance of decomposition, a silent symphony of life and death, is essential for the health and resilience of the Amazon rainforest. It is a constant reminder of the interconnectedness of all living things, the delicate balance that governs the flow of energy and nutrients in this vast, complex ecosystem.

The forest floor, though shrouded in darkness, is a world teeming with life, a testament to the beauty and complexity of nature. It is a reminder that even the smallest organisms, often overlooked, play a vital role in maintaining the delicate balance of life. The next time you walk through the Amazon rainforest, take a moment to appreciate the hidden world beneath your feet, the silent army of decomposers working tirelessly to sustain the life that surrounds you.

The Forest Floor's Importance: A Foundation for Life

The Amazon rainforest, a verdant tapestry woven across the South American landscape, conceals a hidden world beneath its leafy canopy. The forest floor, shrouded in perpetual twilight, may seem like a desolate realm, yet it teems with life, pulsating with unseen activity. It is a complex, dynamic ecosystem, a foundation upon which the entire rainforest rests. .

This seemingly barren expanse, often overlooked in the awe-inspiring grandeur of the towering trees, is a vital component of the Amazon's intricate web of life. It acts as a massive reservoir of nutrients, a crucial link in the forest's intricate food chain, and a vital habitat for an astonishing array of creatures. .

The forest floor, a tapestry woven from fallen leaves, decaying wood, and a carpet of moss and fungi, is a testament to the ceaseless cycle of life and death. This organic matter, decomposing slowly, releases vital nutrients, enriching the soil and providing sustenance for a myriad of organisms. The decomposition process, catalyzed by a symphony of bacteria, fungi, and invertebrates, transforms the dead into the living, feeding the roots of the towering giants above. .

Beneath the tangled undergrowth, an intricate network of roots, interwoven with fungal mycelia, forms a vast underground city. This hidden metropolis, teeming with microscopic life, facilitates the flow of nutrients throughout the forest. The roots of towering trees, reaching deep into the earth, anchor them to the ground and tap into the rich tapestry of nutrients released by decomposing matter. This network of roots, interwoven with fungal mycelia, forms a vast underground city, facilitating the flow of nutrients throughout the forest. .

The forest floor is not a silent stage, but a vibrant theatre of life. A bustling community of decomposers, including earthworms, millipedes, and beetles, diligently work at breaking down fallen leaves and decaying wood. Their tireless labor, unseen and often unnoticed, is the bedrock of the rainforest's productivity, recycling nutrients and transforming them into a form readily accessible by the plants.

The forest floor harbors a staggering diversity of life, from tiny insects to large mammals. The constant moisture and abundance of decaying matter attract a vast array of invertebrates, including ants, termites, and beetles, each playing a crucial role in the intricate tapestry of life. They are the architects of the forest, shaping its structure and composition through their relentless consumption of decaying matter. .

The forest floor provides refuge for numerous animals, from the elusive jaguar to the tiny, vibrantly colored poison dart frog. These creatures rely on the protective cover of the dense undergrowth, the abundance of food sources, and the relative seclusion it offers. The forest floor, a refuge for the shy and the powerful, is a testament to the interconnectedness of all life within the Amazon.

The importance of the forest floor extends beyond its role as a vital habitat. It is a critical element in the regulation of the rainforest's climate. The thick layer of organic matter acts as a sponge, absorbing and retaining moisture, ensuring a steady supply of water for the entire ecosystem. The forest floor also plays a vital role in regulating the flow of water through the rainforest, preventing erosion and maintaining the delicate balance of the ecosystem.

The forest floor, a seemingly overlooked component of the Amazon, is a vital hub of life and a cornerstone of the entire rainforest ecosystem. It is a testament to the interconnectedness of nature, demonstrating how the seemingly insignificant can hold the key to the survival and prosperity of the entire system. Its richness, its diversity, its hidden secrets, make it a captivating realm worthy of exploration and admiration. The forest floor, a reminder that life thrives in the darkest corners, is a compelling testament to the resilience and complexity of nature. .

Chapter 5: The Rivers: Veins of Life

The Amazon River System: A Network of Life and Movement

The River's Impact on the Forest: A Constant Source of Change

The Amazon River, the lifeblood of the rainforest, pulsates with a rhythm that dictates the very essence of the surrounding ecosystem. Its impact on the forest is profound and multifaceted, an intricate tapestry woven from the threads of constant change. The river's ceaseless flow is not merely a conduit for water; it is a dynamic force that shapes the landscape, nourishes the soil, and dictates the distribution of life within the rainforest. This constant source of change is the defining characteristic of the Amazon, its inherent dynamism shaping the evolution of flora and fauna in a perpetual dance of adaptation and survival. .

The river's influence begins with its physical presence, a massive artery carving through the dense vegetation. Its banks, perpetually flooded and receding, create a mosaic of habitats - the "várzea" - which support a unique array of flora and fauna. The annual inundation, a regular cycle of flooding and retreat, acts as a natural fertilizer, depositing rich alluvial soil along the riverbanks. This fertile ground becomes a breeding ground for a vast diversity of plants, many of which are adapted to tolerate the fluctuating water levels. Giant water lilies, their enormous leaves providing a haven for fish and amphibians, flourish in the flooded areas. Tall, majestic trees like the Brazil nut tree, their roots anchored deep within the soil, thrive in the fertile floodplain. .

The river's influence extends far beyond its immediate banks. Its currents, carrying sediment and nutrients, create a dynamic system that extends into the interior of the forest. The deposition of silt along the riverbanks enriches the soil, influencing the distribution and abundance of plant life. This intricate interplay between the river and the forest creates a complex mosaic of vegetation, with species adapted to different levels of nutrient availability and waterlogging. The river's constant flow also helps regulate the forest's microclimate, moderating temperature fluctuations and creating a humid environment conducive to a wide array of life.

The river's influence is further amplified by its intricate connection to the forest's hydrology. The countless tributaries that feed the Amazon River act as drainage channels, ensuring the efficient removal of excess water from the forest floor. This drainage system is crucial for maintaining healthy soil conditions and preventing the stagnation of water, which could lead to the decay of plant matter and the loss of biodiversity. The river also acts as a transport system for essential nutrients, carrying dissolved minerals from the interior of the forest to the floodplain and enriching the soil along its banks.

The movement of water within the Amazon basin is a complex and dynamic process, influenced by factors such as rainfall, evaporation, and the gravitational pull of the earth. The river's flow is not uniform, with periods of high water levels during the rainy season and lower water levels during the dry season. This variability creates a distinct rhythm within the forest, a cycle of flooding and recession that influences the life cycle of many species.

The annual flooding, a defining characteristic of the Amazon, acts as a natural reset button for the forest, rejuvenating the ecosystem and triggering a cascade of ecological events. The receding waters leave behind a rich layer of sediment, providing vital nutrients for plant growth. The flooded areas, now exposed, offer a canvas for a diverse array of life, attracting fish, amphibians, reptiles, and birds seeking food and breeding opportunities. The forest itself undergoes a dramatic transformation during this period, with trees shedding their leaves and fruit to prepare for the dry season.

The river's influence on the forest is further evident in the distribution of its plant and animal life. The "várzea" ecosystems, shaped by the river's annual flooding, harbor a unique array of species adapted to survive the fluctuating water levels. Some trees, like the aguaje palm, develop aerial roots that allow them to breathe even when submerged in water. The leaves of other species, like the giant water lily, act as natural rafts, providing shelter and breeding grounds for fish and amphibians. The river's influence extends beyond the immediate floodplain, shaping the distribution of species throughout the forest. The abundance of insects, attracted to the water's edge, draws in insectivorous birds and mammals, creating a dynamic food web that stretches across the forest.

The river also acts as a natural barrier, influencing the distribution of species across the rainforest. The Amazon River, along with its tributaries, creates a complex network of waterways that divides the forest into distinct regions. This fragmentation can lead to the isolation of populations and the evolution of unique adaptations. The river's presence, therefore, plays a significant role in shaping the genetic diversity of the rainforest, contributing to the incredible biodiversity of the Amazon basin.

The river's influence is not confined to the physical environment; it also plays a vital role in the cultural life of the indigenous communities that inhabit the Amazon. For centuries, indigenous peoples have relied on the river for transportation, fishing, and agriculture. The river's bounty provides a source of sustenance, while its currents carry them to distant villages and trading posts. The river is not just a source of life; it is a vital part of their identity, their history, and their culture.

The river's influence on the forest is a testament to the intricate web of life that exists within the Amazon basin. It is a constant source of change, a dynamic force that shapes the landscape, nourishes the soil, and dictates the distribution of life. The river's influence extends far beyond the immediate floodplain, creating a complex mosaic of ecosystems and contributing to the incredible biodiversity of the Amazon rainforest. The river's presence is a constant reminder of the interconnectedness of all life, a powerful symbol of the resilience and adaptability of nature. .

The Amazon River is more than just a waterway; it is a living entity, a force that shapes the destiny of the rainforest. Its influence is omnipresent, a constant reminder of the interconnectedness of all life. The river's rhythm, its ebb and flow, dictate the very heartbeat of the forest, a constant source of change that drives the evolution of the Amazon's flora and fauna. It is a testament to the power and beauty of nature, a force that sustains life and perpetuates the delicate balance of the ecosystem. As we stand on the banks of the mighty Amazon, we witness the river's influence, a constant source of change that shapes the very essence of the rainforest.

The River's Inhabitants: Adaptations for Aquatic Life

The River's Inhabitants: Adaptations for Aquatic Life.

The Amazon River, a colossal artery coursing through the heart of the rainforest, is a world unto itself. Its waters, teeming with life, are a testament to the extraordinary adaptations that allow creatures to thrive in this unique environment. Here, survival is a constant struggle, a dance between the relentless forces of the river and the ingenuity of its inhabitants. From the microscopic plankton to the colossal Arapaima, each organism has honed remarkable strategies to navigate the challenges and exploit the opportunities presented by this vast aquatic ecosystem.

The river, with its turbulent currents, muddy depths, and fluctuating water levels, presents a dynamic and demanding habitat. To thrive here, organisms must possess adaptations for buoyancy, maneuverability, feeding, and defense. These adaptations are diverse and fascinating, a testament to the incredible diversity of life that the Amazon River supports.

One striking adaptation is the development of specialized body shapes. Fish, the most prominent inhabitants of the river, exhibit an incredible array of forms. The sleek, torpedo-shaped bodies of river dolphins allow them to effortlessly navigate the currents, while the flattened bodies of catfish enable them to maneuver through dense vegetation. The elongated, serpentine bodies of eels allow them to burrow into the soft riverbed, escaping predators and ambushing prey.

The unique hydrodynamic properties of water, its density and viscosity, also play a crucial role in the evolution of adaptations. Many fish have evolved smooth, streamlined bodies, reducing drag and enabling efficient swimming. Others, such as the Piranhas, have developed powerful jaws and teeth, allowing them to overcome the water's resistance and effectively hunt prey.

The river's murky waters, often filled with sediment and debris, pose challenges for visual predators. Some species, like the Electric Eel, have evolved the ability to generate electric fields, allowing them to navigate and hunt in near-total darkness. Others, like the Amazonian Lungfish, have developed unique respiratory systems, enabling them to survive in oxygen-depleted environments.

Adapting to the river's fluctuating water levels is another vital challenge. Many fish, such as the Amazonian Catfish, have developed the ability to breathe air, allowing them to survive in shallow pools during the dry season. Others, like the Amazon River Turtle, have evolved mechanisms for burying themselves in the mud, waiting out periods of low water. .

The river's rich food web supports a diverse array of feeding strategies. Filter feeders, like the Amazonian Paddlefish, use specialized gill rakers to extract tiny organisms from the water. Predators, like the Giant River Otter, have evolved powerful jaws and sharp teeth to hunt fish and other aquatic animals. Detritivores, like the Amazonian Crayfish, play a vital role in the ecosystem, breaking down organic matter and returning nutrients to the river.

Competition for resources is fierce in the Amazon. Fish have evolved a variety of defense mechanisms, including camouflage, mimicry, and venom. Some species, like the Amazonian Poison Dart Frog, produce toxins as a defense mechanism. Others, like the Amazonian Piranha, rely on their aggressive behavior and sharp teeth to deter predators.

The Amazon River, with its diverse inhabitants and their remarkable adaptations, is a living laboratory of evolution. Each species, from the smallest plankton to the largest fish, has honed its own unique set of strategies to survive and thrive in this dynamic and challenging environment. Understanding these adaptations provides valuable insights into the intricate web of life that exists in this vital ecosystem, highlighting the importance of conservation efforts to protect this precious and irreplaceable resource. .

The River's Importance: A Pathway for Nutrients and Life

The Amazon River, the lifeblood of the world's largest rainforest, is more than just a colossal waterway; it is a dynamic artery pulsating with the rhythm of life, transporting essential nutrients and carrying a diverse tapestry of flora and fauna. The river's importance is woven into the very fabric of the rainforest ecosystem, influencing everything from the distribution of plant life to the intricate dance of predator and prey. This intricate interplay makes the Amazon River not merely a conduit for water, but a vital force shaping the very identity of the rainforest.

The Amazon's journey begins high in the Andes Mountains, where melting glaciers and torrential rains feed the headwaters. As the river descends, its waters carry a wealth of nutrients, eroded from the Andean slopes and deposited along its course. These nutrients, primarily nitrogen, phosphorus, and silica, are essential for the growth and survival of the rainforest's vibrant flora. As the river flows across the vast Amazon basin, it acts as a lifeline for the surrounding vegetation, distributing these vital elements far and wide. This nutrient-rich water, flooding the rainforest during the wet season, revitalizes the soil and creates a fertile environment for a staggering array of plants, from towering trees to delicate orchids.

The Amazon River's significance extends far beyond plant life, playing a crucial role in the survival of the rainforest's diverse animal communities. The river provides a habitat for a vast array of aquatic species, from the iconic pink river dolphins to the elusive electric eel. Its waters are a highway for migrating fish, allowing them to travel thousands of kilometers in search of food and breeding grounds. These fish, in turn, provide a vital food source for many terrestrial animals, including jaguars, tapirs, and countless bird species. The river's influence extends beyond its direct borders, shaping the lives of animals living far from its shores.

The Amazon's constant movement creates a dynamic environment, fostering a constant exchange of nutrients and life. The river's ebb and flow influence the distribution of plant life, carving out unique habitats along its banks. Floodplains, inundated during the wet season, provide a rich, nutrient-rich environment for the growth of aquatic plants, creating havens for fish and amphibians. During the dry season, the river recedes, exposing these floodplains, allowing terrestrial animals to exploit the rich food resources they offer. This intricate cycle of flooding and recession ensures the rainforest's ecological balance, supporting a vast array of life.

The river's role in the rainforest's intricate food web is further amplified by its ability to transport organic matter. As leaves, branches, and other organic debris fall into the river, they are broken down by bacteria and fungi, releasing essential nutrients that fuel the growth of phytoplankton, the base of the aquatic food chain. This organic matter, along with the nutrients carried from the Andes, provides sustenance for a vast array of organisms, from microscopic plankton to large, predatory fish. The river, therefore, acts as a critical link between the terrestrial and aquatic ecosystems, facilitating the transfer of energy and nutrients across the rainforest landscape.

The Amazon River's influence extends even beyond the boundaries of the rainforest, reaching far into the Atlantic Ocean. Its vast discharge of freshwater and sediment significantly alters the ocean's salinity and nutrient levels, impacting marine ecosystems. The river's plume, a vast cloud of freshwater and sediment, supports a rich diversity of marine life, including the critically endangered Amazon River dolphin. This intricate web of interactions highlights the Amazon's global significance, underscoring the interconnectedness of ecosystems and the profound impact a single waterway can have on the planet's natural systems.

The Amazon River, therefore, is not simply a body of water; it is a vibrant ecosystem, a dynamic force shaping the rainforest's landscape, influencing the distribution of flora and fauna, and providing sustenance for an astounding array of life. Its importance extends beyond the rainforest, influencing marine ecosystems and highlighting the interconnectedness of our planet's natural systems. The river's legacy transcends the physical environment, embodying the very essence of life, pulsing with the rhythm of the rainforest, and reminding us of the vital role waterways play in the intricate tapestry of life on Earth. .

Chapter 6: The Trees: Pillars of the Forest

The Diversity of Amazonian Trees: A Wide Range of Adaptations

The Diversity of Amazonian Trees: A Wide Range of Adaptations.

The Amazon rainforest, a sprawling tapestry of life, thrives on an astonishing array of trees, each a testament to the power of adaptation. These arboreal giants, in their remarkable diversity, have carved niches within the rainforest ecosystem, showcasing a spectrum of adaptations that allows them to flourish in the unique challenges of this environment. .

The sheer density of the canopy, a defining feature of the rainforest, presents a primary challenge – access to sunlight. Trees have developed an array of strategies to overcome this hurdle. Some, such as emergent trees, like the towering Brazil nut tree (Bertholletia excelsa), strive for the highest reaches, breaching the canopy to bathe in unfiltered sunlight. Their smooth, barkless trunks minimize epiphyte growth, ensuring maximum sunlight absorption. Others, like the kapok tree (Ceiba pentandra), possess expansive, buttress roots that anchor them firmly against the torrential rains, while their massive canopies provide shade for the understory plants. .

The Amazon's diverse soil types, ranging from nutrient-poor white sands to fertile terra firme, pose another challenge. Trees have evolved specialized adaptations to thrive in these contrasting environments. Species like the Mauritia flexuosa, the majestic moriche palm, flourish in seasonally inundated areas, their roots capable of extracting nutrients from waterlogged soils. Others, like the towering mahogany tree (Swietenia macrophylla), have developed deep taproots that delve into the earth, reaching nutrient-rich layers. .

The constant battle against herbivory shapes the Amazonian trees. Defenses against leaf-eating insects and browsing animals are a constant evolutionary arms race. Many species, like the rubber tree (Hevea brasiliensis), produce latex, a sticky, toxic substance that deters herbivores. Others, like the inga tree (Inga spp.), bear thorns or spines, offering physical protection. .

The Amazonian trees are not solitary giants. They engage in complex ecological interactions, forging partnerships with fungi, microbes, and animals. Mycorrhizae, symbiotic fungi, provide essential nutrients to trees while receiving carbohydrates in return. Ants, like the Azteca species, form symbiotic relationships with certain trees, providing protection against herbivores and competitors in exchange for shelter and food. The Amazon's vibrant ecosystem thrives on these intricate relationships, a testament to the interconnectedness of life in the rainforest.

The Amazonian trees are not just pillars of the forest; they are architects of its complex web of life. Their diversity, a product of millions of years of adaptation, forms the foundation of the rainforest's extraordinary biodiversity. Their survival hinges on a delicate balance, where every adaptation is honed by the forces of nature, showcasing the resilience and ingenuity of life in the Amazon. .

The Tree's Role in the Ecosystem: Providing Structure and Shelter

In the heart of the Amazon, where life teems with vibrant complexity, the mighty trees stand as the very pillars of the forest. They are not just passive giants, but active participants in the intricate web of life, providing the very structure and shelter upon which countless species depend. Their towering canopies form a verdant ceiling, filtering sunlight and creating microclimates that nourish a diverse array of flora and fauna. .

The trees themselves are not monolithic structures; they are ecosystems in their own right. Each one harbors a community of organisms, from the tiny insects that inhabit the crevices of their bark to the epiphytes that cling to their branches, creating a vertical tapestry of life. This intricate tapestry begins with the roots, anchoring the tree firmly to the earth and acting as a conduit for water and nutrients. .

The roots not only sustain the tree but also influence the very soil composition, binding it together and preventing erosion. This stability is critical for the entire forest ecosystem, as the soil provides the foundation for countless other plants, and the trees, with their root systems, ensure that this foundation remains intact. .

Ascending the trunk, one encounters a world of microhabitats. The rough bark provides shelter for a myriad of insects, amphibians, and reptiles, each adapted to its specific niche. This diversity is further enhanced by the presence of mosses, lichens, and fungi, which thrive in the damp environment provided by the bark's surface. .

Finally, we reach the crown, the pinnacle of the tree's structure. Here, the leaves form a dense canopy, filtering sunlight and creating a mosaic of light and shade that allows a diverse array of plant life to thrive on the forest floor. The leaves themselves are also a valuable resource, providing food for countless herbivores, from insects to sloths to monkeys. .

The canopy is not only a source of food but also a vital refuge for many animals. The dense foliage provides shelter from predators and the elements, allowing birds to build nests, monkeys to swing through the branches, and bats to roost in the safety of the tree's embrace. The branches also serve as highways, connecting different parts of the forest and facilitating the movement of animals. .

But the trees are not simply providers of shelter and resources; they are also active participants in the forest's intricate web of relationships. Their leaves release oxygen into the atmosphere, contributing to the delicate balance of gases that sustains life on Earth. They also absorb carbon dioxide, helping to mitigate the effects of climate change. .

Furthermore, the trees' presence influences the forest's microclimate. Their canopies modify temperature and humidity, creating a more stable environment that allows for a greater diversity of plant and animal life. The dense foliage also reduces wind speed, protecting the forest floor from the harshest elements. .

Beyond their individual contributions, the trees work together to form a vast interconnected network. Their roots intertwine, creating a subterranean web that allows them to share water and nutrients. Their leaves form a continuous canopy, filtering sunlight and creating a uniform environment that supports a vast array of life. .

As the sun sets, casting long shadows through the forest, the trees continue their silent yet vital work. They provide refuge to nocturnal creatures, their branches offering a safe haven from the perils of the night. Their leaves, rustling in the wind, create a symphony of sound, a subtle reminder of the interconnectedness of life within the forest. .

The trees of the Amazon are not just passive giants, they are the very backbone of the forest, providing structure, shelter, and a framework for the complex web of life that thrives within its depths. Their presence is a testament to the resilience and beauty of nature, a silent force that sustains an entire ecosystem. .

As we navigate the Amazon's labyrinthine paths, we are constantly reminded of the vital role that trees play in this complex and dynamic environment. They are the architects of the forest, shaping its very structure and influencing the life that thrives within it. Their presence is a constant reminder of the interconnectedness of all living things, a testament to the delicate balance of nature that we must strive to preserve. .

The Importance of Trees in the Carbon Cycle: Regulating Climate

The Amazon rainforest, a sprawling tapestry of emerald green, stands as a monument to the intricate dance between life and the planet's climate. It is here, in this vast expanse of vibrant ecosystems, that the role of trees in the carbon cycle takes center stage, a silent and often overlooked drama of global proportions. Trees, those towering giants of the forest, are not mere passive bystanders in this planetary ballet but rather the very choreographers, directing the flow of carbon and influencing the global climate in ways that are only beginning to be understood.

The Amazon, with its dizzying array of plant and animal life, is a vibrant testament to the fundamental role of trees in regulating the Earth's climate. These arboreal giants, through a complex and interconnected series of processes, act as the lungs of the planet, breathing in carbon dioxide and releasing oxygen. The process of photosynthesis, the very lifeblood of these towering sentinels, forms the heart of their role in the carbon cycle. Through their leaves, they absorb carbon dioxide from the atmosphere, transforming it into sugars, the fuel that powers their growth. This stored carbon, the very essence of their being, is then incorporated into their wood, their branches, and their roots, becoming a vital part of the forest's structure.

But the role of trees in the carbon cycle extends far beyond the simple act of photosynthesis. Their intricate root systems, sprawling beneath the forest floor, act as a massive carbon sink, capturing and storing vast amounts of carbon from the atmosphere. This process, known as carbon sequestration, is a cornerstone of the Earth's climate regulation, acting as a natural buffer against the escalating greenhouse effect. The Amazon rainforest, with its staggering density of trees, serves as a colossal carbon reservoir, holding more carbon than any other terrestrial ecosystem on the planet.

The ability of trees to sequester carbon is not a static phenomenon but a dynamic process that is intimately tied to the forest's ecosystem. Factors such as rainfall, temperature, and soil fertility all influence the rate at which trees capture and store carbon. Moreover, the age of the forest plays a crucial role, with older, mature forests exhibiting significantly higher carbon sequestration rates than younger, regenerating forests. This intricate interplay of factors highlights the vital importance of maintaining the integrity of existing forests, ensuring their continued ability to absorb and store carbon.

The intricate relationship between trees and the carbon cycle extends beyond the forest floor, reaching into the very heart of the atmosphere. As trees age and eventually die, their accumulated carbon is released back into the atmosphere through decomposition. This process, however, is not a simple release of carbon but rather a complex series of interactions involving fungi, bacteria, and other organisms. These decomposers play a vital role in breaking down the deadwood, releasing carbon back into the atmosphere, but also sequestering a portion of it within the soil, creating a long-term carbon sink.

The delicate balance between carbon sequestration and release, between life and death, is crucial for maintaining a stable climate. The Amazon rainforest, with its rich biodiversity and complex web of interactions, exemplifies this delicate balance. The forest's ability to act as a net carbon sink, absorbing more carbon than it releases, is a testament to the critical role of trees in regulating global climate. However, this balance is increasingly under threat, as deforestation and degradation threaten to disrupt the natural carbon cycle.

Deforestation, the rampant clearing of forests for agriculture, logging, and other human activities, has devastating consequences for the carbon cycle. By removing trees, we are not only eliminating a vital carbon sink but also releasing stored carbon back into the atmosphere, further exacerbating the greenhouse effect. The burning of forests, often used to clear land for agriculture, further exacerbates the problem, releasing vast amounts of carbon dioxide into the atmosphere, accelerating the rate of global warming.

The destruction of the Amazon rainforest, with its staggering carbon storage capacity, is a stark reminder of the interconnectedness of our planet's systems. The loss of this vital carbon sink not only has immediate consequences for the local climate but also contributes to global climate change, threatening the stability of the entire planet. The fight to preserve the Amazon rainforest, to protect its vast carbon reservoirs, is a fight for our very future.

The Amazon's carbon story, however, is not one of despair but rather a call to action. The resilience of the rainforest, its ability to regenerate and recover, offers a glimmer of hope. The planting of trees, reforestation efforts, and sustainable forest management practices are all crucial steps in mitigating the effects of climate change. By investing in the health and well-being of the forest, we are investing in the future of our planet.

The trees of the Amazon, those silent sentinels of the forest, are a stark reminder of the interconnectedness of life on Earth. Their role in the carbon cycle, their ability to regulate climate, is a testament to the vital importance of preserving and protecting our planet's natural resources. The future of the Amazon, and indeed the future of our planet, lies in recognizing the profound value of these towering giants and embracing sustainable practices that ensure their continued existence. For it is in the green heart of the Amazon, in the intricate dance of life and climate, that we find the key to a sustainable future.
.

Threats to Amazonian Trees: Deforestation and Climate Change

The Amazon rainforest, a tapestry of vibrant life, is a testament to the resilience of nature. However, its verdant expanse is under siege, facing a dual threat that looms large over its very existence: deforestation and climate change. The trees, the very backbone of this ecosystem, are bearing the brunt of these assaults, their survival hanging precariously in the balance. .

Deforestation, the relentless clearing of forest land for agriculture, logging, and infrastructure development, is the most immediate and visible threat. The insatiable demand for agricultural commodities like soy, beef, and palm oil fuels the expansion of land use, swallowing vast tracts of pristine forest. Logging operations, driven by the global timber trade, leave behind scars on the landscape, extracting valuable wood at the cost of ecological stability. .

The consequences of deforestation are far-reaching and devastating. The loss of trees disrupts the delicate balance of the Amazonian ecosystem. It disrupts the intricate web of life, impacting biodiversity and ecosystem services. The destruction of habitat jeopardizes the survival of countless species, driving them towards extinction. Furthermore, the removal of trees disrupts the natural carbon cycle, releasing vast quantities of carbon dioxide into the atmosphere, exacerbating the effects of climate change.

Climate change, a global phenomenon driven by human activity, casts a long shadow over the Amazon. Rising global temperatures, coupled with altered precipitation patterns, are impacting the forest in profound ways. The increased frequency and intensity of droughts are stressing the trees, making them more vulnerable to diseases, pests, and fire. The loss of moisture further amplifies the risk of wildfires, turning the once-lush rainforest into a landscape ravaged by flames.

The effects of climate change on the Amazonian trees are multifaceted. Rising temperatures and changes in precipitation patterns disrupt the delicate balance of the rainforest ecosystem. The trees, adapted to specific climatic conditions, are forced to cope with altered temperatures, rainfall, and humidity levels. This shift can lead to stress, reduced growth rates, and increased vulnerability to disease and pests. .

The interconnectedness of the rainforest ecosystem is evident in the impact of climate change on its various components. Changes in temperature and precipitation patterns can disrupt the complex interactions between trees, fungi, insects, and other organisms, leading to cascading effects throughout the ecosystem. The loss of keystone species, such as pollinators and seed dispersers, can further destabilize the forest, contributing to the decline of tree populations.

The fate of the Amazonian trees is inextricably linked to the fate of the entire rainforest ecosystem. The loss of these towering giants would be a monumental tragedy, impacting not only the biodiversity of the Amazon but also the global climate. The Amazon plays a vital role in regulating the global carbon cycle, absorbing vast amounts of carbon dioxide from the atmosphere. The continued loss of trees would result in the release of significant amounts of carbon dioxide, accelerating the pace of climate change.

The preservation of the Amazonian trees is a crucial task for humanity. It requires a multifaceted approach, encompassing measures to curb deforestation, mitigate climate change, and promote sustainable land management practices. International cooperation, national policies, and community-based initiatives are all essential components of this complex endeavor. .

The Amazonian trees stand as sentinels of a delicate ecosystem, a testament to the power and beauty of nature. Their survival is a matter of urgency, requiring immediate action to address the threats posed by deforestation and climate change. The responsibility to protect these vital pillars of the forest rests on the shoulders of all humankind. The fate of the Amazon, and indeed the fate of the planet, hangs in the balance. .

Chapter 7: The Birds: Symphony of the Rainforest

The Amazon's Avian Diversity: A Colorful Spectacle

The Amazon's Avian Diversity: A Colorful Spectacle.

The Amazon rainforest, a sprawling tapestry of emerald green, teems with life. But amidst the cacophony of rustling leaves and the symphony of insects, one sound reigns supreme: the chorus of birdsong. The Amazon boasts an avian diversity unmatched anywhere else on the planet, a breathtaking spectacle of color, sound, and ecological adaptation. This avian tapestry is woven with threads of vibrant plumage, intricate songs, and specialized niches, each thread contributing to the complex and fascinating web of life in the rainforest.

The Amazon's avian diversity is a consequence of its immense size, geographical isolation, and unparalleled biodiversity. This ecological richness has fostered the evolution of thousands of bird species, each uniquely adapted to its specific environment. From the majestic harpy eagle soaring high above the canopy to the tiny hummingbird darting between flowering vines, the Amazon's avian community showcases an astonishing array of adaptations, each a testament to the power of natural selection.

The sheer number of species is staggering. Scientists estimate that over 1,600 bird species inhabit the Amazon Basin, accounting for roughly 10% of all known bird species worldwide. This impressive figure is a consequence of the rainforest's diverse habitats, ranging from flooded forests and terra firme to savannas and cloud forests. Each habitat niche supports a unique community of birds, each species adapted to exploit specific food sources and ecological opportunities.

The Amazon's bird diversity is not just about numbers; it's about the spectacle of color and sound that fills the air. Imagine a canopy alive with the vibrant hues of scarlet macaws, emerald toucans, and the turquoise plumage of a hummingbird hovering at a flower. The air hums with the rhythmic drumming of woodpeckers, the sweet melodies of tanagers, and the guttural calls of parrots. This symphony of sound and color is an intrinsic part of the Amazon's natural beauty, a breathtaking testament to the wonders of evolution.

The Amazon's avian diversity is more than just a captivating spectacle; it plays a crucial role in the rainforest's delicate ecosystem. Many bird species serve as vital seed dispersers, contributing to the regeneration of the rainforest by transporting seeds to new locations. Others act as pollinators, ensuring the reproductive success of flowering plants. Some species are specialized predators, regulating populations of insects and small vertebrates, thereby maintaining the ecological balance of the rainforest.

The Amazon's avian diversity is also a source of cultural inspiration and economic value. Many indigenous communities rely on birds for food, feathers for ornamentation, and song for spiritual rituals. Birdwatching tourism, a growing industry in the region, provides economic benefits and encourages conservation efforts. However, this fragile ecosystem faces numerous threats, including deforestation, habitat fragmentation, and illegal wildlife trade.

Understanding the Amazon's avian diversity is not just about appreciating its beauty but also recognizing its ecological significance and its vulnerability. As we continue to unravel the mysteries of this remarkable ecosystem, we must strive to protect its vibrant avian tapestry, ensuring that the symphony of the rainforest continues to resonate for generations to come. .

Adaptations for Flight and Survival: A World of Specializations

The Amazon rainforest, a symphony of life, is a tapestry woven with the vibrant hues of countless bird species. This verdant world, a labyrinth of towering trees and shimmering waterways, has birthed a breathtaking diversity of avian life, each species meticulously sculpted by evolution to thrive within this intricate ecosystem. Here, adaptations for flight and survival are not mere feats of nature; they are exquisite masterpieces, each tailored to the specific challenges of this remarkable environment.

Take, for instance, the toucans, their massive beaks a testament to their frugivorous lifestyle. These beaks, seemingly cumbersome, are lightweight marvels of structural ingenuity, allowing them to reach and pluck fruit with effortless precision. Their bright, contrasting colors, a symphony of blues, reds, and yellows, are not mere ornamentation; they serve as powerful signals, communicating dominance, attracting mates, and even warning potential predators.

The hummingbirds, aerial acrobats of the rainforest, represent another exquisite adaptation. Their iridescent plumage, shimmering with an almost otherworldly brilliance, reflects sunlight in dazzling displays, attracting mates and dazzling rivals. Their wings, a blur of motion, are capable of hovering, darting, and even flying backwards, maneuvering with an agility that defies gravity. Their long, slender beaks, often curved and specialized for specific flowers, allow them to access nectar, fueling their astonishing energy expenditure.

Within the forest canopy, the parrots, masters of mimicry and social interaction, reveal a unique blend of adaptations. Their vibrant plumage, a kaleidoscope of colors, serves as a visual language, communicating complex messages within their flocks. Their powerful beaks, adept at cracking seeds and nuts, provide them with a diverse diet, enabling them to thrive in a world of fluctuating resources. Their vocalizations, a symphony of screeches, whistles, and calls, are powerful tools for communication, establishing territories, and attracting mates.

The raptors, apex predators of the avian world, embody the pinnacle of predatory adaptations. Their keen eyesight, capable of spotting prey from dizzying heights, allows them to dominate the skies. Their talons, razor-sharp and powerful, are instruments of death, capable of delivering swift and efficient kills. Their flight, silent and precise, embodies the ultimate mastery of aerial navigation, enabling them to surprise their prey and escape with their bounty.

The kingfishers, masters of the aquatic realm, showcase adaptations tailored for a life spent in and around water. Their bright plumage, often adorned with contrasting colors, serves as a signal of their predatory prowess, warning rivals and attracting mates. Their sharp beaks, designed for spearing fish with precision, allow them to snatch their prey from the depths of rivers and streams. Their flight, a combination of powerful dives and graceful glides, enables them to navigate the intricate landscape of the rainforest canopy and the water's edge with ease.

The tanagers, a vibrant tapestry of colors adorning the rainforest canopy, embody the artistry of avian adaptation. Their intricate plumage, a symphony of reds, yellows, blues, and greens, serves as a dazzling display of courtship, attracting mates and establishing social hierarchies. Their melodic songs, a chorus of whistles and trills, are used to communicate with each other, defend territories, and attract potential partners.

Beyond the canopy, the cuckoos, masters of brood parasitism, showcase a unique adaptation for survival. Their glossy plumage, often blending seamlessly with their surroundings, allows them to remain hidden as they sneak into the nests of other birds, laying their eggs amongst the host's clutch. Their chicks, with their powerful beaks and relentless hunger, outcompete the host's offspring for food and resources, ensuring their own survival.

The Amazon rainforest, with its breathtaking diversity, presents a myriad of challenges for avian life. Each species, through the intricate dance of evolution, has honed its adaptations, shaping its physique, behavior, and survival strategies to thrive in this demanding environment. From the vibrant hues of the toucans to the masterful mimicry of the parrots, from the aerial prowess of the hummingbirds to the silent efficiency of the raptors, each bird tells a story of adaptation and survival, painting a vivid tableau of life's resilience and ingenuity.

The Birds' Role in the Ecosystem: Seed Dispersal and Pollination

The Birds' Role in the Ecosystem: Seed Dispersal and Pollination.

The Amazon rainforest, a symphony of life, reverberates with the songs and calls of a vibrant avian community. Beyond their captivating melodies, these feathered denizens play a pivotal role in the intricate web of life that sustains the rainforest, acting as vital agents of seed dispersal and pollination. The intricate dance between birds and plants is a testament to the remarkable interconnectedness of nature, showcasing a delicate balance that ensures the survival and propagation of countless species.

Seed Dispersal: A Flight of Life.

Imagine a tiny seed, nestled deep within a fleshy fruit, its destiny seemingly confined to the forest floor. But fate, in the form of a vibrant toucan, intervenes.

Attracted by the fruit's vibrant hues and sweet aroma, the toucan plucks the fruit, its powerful beak effortlessly cracking the outer shell. The seed, freed from its confines, embarks on an unexpected journey, carried aloft by the toucan's flight.

This scenario, a common occurrence in the Amazon, highlights the crucial role birds play in seed dispersal. By consuming fruits and excreting undigested seeds, birds act as potent agents of plant propagation, transporting seeds far beyond the reach of the parent plant. This dispersal strategy, known as endozoochory, is particularly important in the dense, complex rainforest environment, where seeds are often challenged to reach suitable germination sites.

The effectiveness of seed dispersal by birds is magnified by the sheer diversity of avian species, each adapted to consume and disperse specific types of fruit. The toucan, with its robust beak, specializes in cracking open large, hard-shelled fruits, while smaller birds, like tanagers and manakins, favor softer, berry-like fruits. This diverse dietary spectrum allows birds to disperse a wide range of seeds, facilitating the establishment and maintenance of plant communities across the rainforest.

Furthermore, birds contribute to seed dispersal through their ability to fly long distances. This allows them to transport seeds to distant locations, expanding the geographic range of plants and promoting genetic diversity. This dispersal, coupled with the selective pressure of seed ingestion and excretion, ensures that viable seeds are deposited in suitable environments, increasing the likelihood of germination and seedling establishment.

Pollination: The Dance of Life.

The rainforest teems with vibrant flowers, each a beacon of color and fragrance, attracting pollinators that play a crucial role in their reproductive success. Among these pollinators, birds are prominent, their beaks and bodies often adapted to specific flower types, creating a captivating dance of coevolution.

Hummingbirds, with their long, slender beaks and hovering flight, are particularly adept at pollinating tubular flowers. Their bodies, coated in pollen as they feed on nectar, act as efficient pollen carriers, transferring the precious genetic material from flower to flower.

Other bird species, like tanagers and honeycreepers, possess specialized beaks designed to access the nectar of specific flowers. These adaptations ensure efficient pollination, minimizing pollen loss and maximizing the chances of successful fertilization.

The pollination process, facilitated by the intricate relationship between birds and flowers, is crucial for the survival and perpetuation of countless rainforest plant species. Without the vital role of avian pollinators, the vibrant tapestry of the rainforest would be significantly diminished, leading to a cascade of ecological consequences.

Beyond the Symphony: The Interconnected Web.

The roles of seed dispersal and pollination are not merely isolated events but rather intricately interwoven threads in the tapestry of the rainforest ecosystem. The actions of birds, seemingly mundane acts of feeding and foraging, have profound implications for the health and resilience of the rainforest.

Seed dispersal, by facilitating plant colonization and promoting genetic diversity, contributes to the overall biodiversity of the rainforest. Pollination, by ensuring the reproductive success of countless plant species, maintains the intricate food web that sustains the rainforest ecosystem. .

These interconnected processes, orchestrated by the diverse avian community, highlight the delicate balance of nature, a symphony of life where each player contributes to the grand composition. Understanding the intricate roles of birds in seed dispersal and pollination is crucial for conservation efforts, safeguarding the delicate equilibrium of this vital ecosystem.

As we delve deeper into the Amazon rainforest, we are struck by the remarkable interconnectedness of life. The seemingly simple act of a bird consuming a fruit or sipping nectar has profound implications for the survival and propagation of countless species. These avian ambassadors, through their unique adaptations and foraging behaviors, ensure the continued flourishing of the rainforest, a vital treasure trove of biodiversity that must be cherished and protected for generations to come.

Birds in Peril: Threats to Amazonian Bird Populations

Birds in Peril: Threats to Amazonian Bird Populations.

The Amazon rainforest, a sprawling tapestry of verdant life, teems with a symphony of avian voices. From the vibrant plumage of the toucans to the ethereal calls of the hoatzin, the Amazon's avian diversity is a testament to the unparalleled richness of this ecosystem. However, this vibrant symphony is facing a silent crisis – a growing list of threats that imperil the very existence of Amazonian bird populations. .

Habitat loss, driven by the relentless march of deforestation and agricultural expansion, poses the most significant threat. As the rainforest shrinks, so too do the territories of countless bird species. This not only reduces the availability of food and nesting sites but also disrupts the delicate balance of the ecosystem, leading to cascading effects throughout the food chain.

Beyond outright habitat destruction, the fragmentation of the rainforest into isolated patches poses a serious challenge. Birds reliant on large, contiguous forest areas struggle to thrive in fragmented landscapes, facing increased vulnerability to predation and reduced access to critical resources. .

Climate change, a looming shadow over the entire planet, adds another layer of complexity to the threats facing Amazonian birds. Rising temperatures, altered rainfall patterns, and increased frequency of extreme weather events disrupt established ecological relationships and push many species to the brink. For instance, changes in rainfall patterns can impact the availability of food sources for certain birds, while extreme heat events can lead to breeding failures and even mortality. .

The illegal wildlife trade further exacerbates the plight of Amazonian birds. Many species, prized for their stunning plumage or vocalizations, are captured and traded, often destined for the pet trade or for use in traditional medicine. This practice not only removes individuals from the wild but also disrupts the delicate balance of populations, contributing to long-term declines. .

Pollution, a silent killer, further adds to the woes of Amazonian birds. Agricultural runoff, industrial waste, and oil spills contaminate water sources, impacting the health of birds and their food sources. .

These multifaceted threats are not only impacting individual species but also contributing to a decline in the overall biodiversity of the Amazonian avifauna. The loss of these birds not only diminishes the beauty and wonder of the rainforest but also has broader implications for the entire ecosystem. .

Birds play crucial roles in seed dispersal, pollination, and pest control. Their disappearance disrupts these critical ecosystem services, leading to cascading effects on plant communities and the overall health of the rainforest. .

Recognizing the gravity of the situation, conservation efforts are crucial to mitigating the threats facing Amazonian birds. Protecting remaining forest habitat through the establishment of national parks and protected areas is paramount. Sustainable forest management practices can help balance human needs with the conservation of biodiversity. .

Combating the illegal wildlife trade through stricter regulations and enforcement is essential. Addressing climate change through international cooperation and promoting sustainable practices are crucial for ensuring the long-term survival of Amazonian birds. .

The future of Amazonian birds hangs in the balance, dependent on our collective efforts to address the threats they face. It is a stark reminder that the health and well-being of our planet are inextricably linked to the fate of even the smallest creatures, and that the symphony of the rainforest, once threatened, may forever be silenced. .

Chapter 8: The Mammals: Giants and Shadows

The Amazon's Mammalian Diversity: A Range of Shapes and Sizes

The Amazon's Mammalian Diversity: A Range of Shapes and Sizes .

The Amazon, a tapestry woven with emerald green leaves and the constant hum of life, harbors a mammalian diversity unmatched anywhere else on the planet. This vast expanse of rainforest, spanning nine countries and covering nearly 40% of South America, offers a mesmerizing array of habitats, each teeming with unique and fascinating creatures. From the towering canopy, where acrobatic primates swing through the branches, to the murky depths of the rivers, where sleek otters glide effortlessly, the Amazon's mammalian cast is a testament to the power of evolution and adaptation. .

Within this remarkable assemblage of life, size plays a pivotal role, painting a vibrant picture of ecological niches and the intricate dance of predator and prey. The giants, such as the majestic jaguar, the Amazon river dolphin, and the imposing tapir, dominate the landscape, wielding strength and power. They are apex predators, maintaining the delicate balance of their ecosystems through their influence on prey populations. But the Amazon's grandeur is also defined by its miniature denizens, the tiny opossums, the agile marmosets, and the elusive rodents, each playing an essential role in the intricate web of life. .

The diversity of shapes and sizes reflects the incredible range of adaptations that have allowed these mammals to thrive in the Amazon's diverse environments. The slender limbs of the spider monkey, uniquely suited for navigating the canopy, contrast sharply with the powerful legs of the tapir, designed for navigating dense undergrowth. The sharp claws of the jaguar, a testament to its predatory prowess, stand in stark contrast to the sleek, paddle-like tails of the river dolphins, perfect for maneuvering through the murky waters. These morphological variations are not mere aesthetic flourishes; they are the very essence of survival, honed through millennia of evolution to exploit the specific resources and challenges presented by their unique habitats.

However, the Amazon's mammalian tapestry is far from static. The delicate balance of this dynamic ecosystem is constantly challenged by the forces of human encroachment. Deforestation, hunting, and the introduction of invasive species threaten to unravel the intricate web of life that has sustained these incredible creatures for centuries. The Amazon's iconic giants, from the jaguar to the giant river otter, are particularly vulnerable to these threats, their populations dwindling as their habitats shrink and their prey becomes scarce. .

The loss of these majestic creatures would not only be a tragic blow to biodiversity but would also have ripple effects throughout the entire Amazonian ecosystem. The decline of apex predators like the jaguar can lead to a cascade effect, disrupting the balance of prey populations and potentially even altering the composition of plant communities. This underscores the interconnectedness of life in the Amazon, where every creature plays a crucial role in maintaining the delicate equilibrium. .

But hope still remains. Conservation efforts, driven by passionate individuals and organizations, are working tirelessly to protect the Amazon's mammalian diversity. The establishment of protected areas, the implementation of sustainable forestry practices, and the fight against illegal hunting are crucial steps towards safeguarding the future of these remarkable creatures. .

As we delve deeper into the Amazon's mammalian tapestry, we encounter not only the magnificence of size but also the elegance of adaptation. The diverse forms, sizes, and behaviors of these remarkable animals paint a vibrant picture of the interconnectedness of life, reminding us of the delicate balance of nature and the vital importance of conservation in preserving the Amazon's extraordinary biodiversity for future generations. .

Adaptations for Survival in the Jungle: From Stealth to Strength

The Amazon, a realm of vibrant green, teeming with life, is a canvas upon which evolution has painted a spectacular array of adaptations. Nowhere is this more evident than in the mammal kingdom, where giants and shadows alike have carved niches within this intricate web of survival.

The jaguar, the apex predator of the Amazon, embodies the very essence of stealth and strength. Its tawny coat, dappled with black rosettes, provides camouflage amidst the dappled sunlight filtering through the canopy. Its powerful build, honed by centuries of hunting, allows it to stalk through dense undergrowth, its muscular limbs silent on the forest floor. The jaguar's eyes, adapted for nocturnal vision, pierce the darkness, granting it a distinct advantage in the low light conditions of the jungle. Its jaws, armed with formidable canines, deliver a crushing bite capable of bringing down even the largest prey. This symphony of adaptations makes the jaguar a formidable hunter, a master of its domain, striking from the shadows with deadly precision.

Yet, the Amazon is not solely a stage for the titans. Smaller mammals, like the agouti, have developed their own ingenious strategies to navigate this complex ecosystem. With their long, slender legs, these rodents can dart through the undergrowth with remarkable speed, evading the watchful eyes of predators. Their sharp incisors, constantly growing, are perfect for cracking open tough nuts and seeds, their primary food source. Their keen sense of smell, honed over generations, allows them to locate these hidden treasures, even beneath the dense layers of leaves. The agouti's agility, coupled with its dietary adaptability, makes it a survivor, a testament to the resilience of nature's smallest creatures.

Survival in the Amazon is not merely about brute strength or stealth. It is also about the ability to adapt to the unique challenges of a fluctuating environment. The howler monkey, for example, has developed a unique vocalization strategy to communicate across vast distances. Their distinctive calls, resonating through the dense foliage, can be heard for miles, serving as a means of territorial defense, mate attraction, and coordination within their social groups. This vocal prowess allows them to maintain a strong social fabric, vital for navigating the complex dynamics of the Amazonian ecosystem.

The Amazon's aquatic realm is no less complex. The river dolphin, a creature of graceful movements and enigmatic behavior, has adapted to a life entirely submerged in the Amazon's murky waters. Its sleek, torpedo-shaped body, devoid of any dorsal fin, allows it to navigate through dense vegetation with ease. Its specialized echolocation system, honed by countless generations, grants it a detailed sonar image of its surroundings, enabling it to hunt fish and avoid obstacles in the murky depths. The river dolphin's unique adaptations highlight the remarkable versatility of life, showcasing the ability to thrive even in the most demanding environments.

The Amazon's biodiversity extends even to the smallest of its inhabitants. The pygmy marmoset, the smallest monkey in the world, has evolved a unique strategy for navigating the rainforest canopy. Its tiny size, coupled with its agile limbs and prehensile tail, allows it to move with remarkable speed and precision through the branches, accessing food sources that larger primates cannot reach. Its diet, primarily consisting of insects and sap, further underscores its adaptability, enabling it to thrive in a niche that few other creatures can occupy.

The Amazon, a world of endless green, pulsates with life, each species a testament to the power of adaptation. From the majestic jaguar to the diminutive pygmy marmoset, each creature has carved its own path, its own unique strategy for survival within this intricate tapestry of life. The Amazon is a living laboratory of evolution, a testament to the remarkable diversity and resilience of nature, where giants and shadows alike dance to the rhythm of survival.

The Mammals' Role in the Ecosystem: Predators, Herbivores, and More

The Amazon, a tapestry woven with emerald green and shadowed by a canopy of ancient giants, teems with life. It is a symphony of chirps, rustles, and roars, and at the heart of this vibrant orchestra, the mammals play a crucial role. They are the architects of the rainforest's intricate web of life, shaping its structure and dynamism in ways both subtle and profound. Their existence is a testament to the power of adaptation, a spectacle of diversity that showcases nature's boundless creativity. .

The most prominent players in this mammalian drama are the predators, the apex hunters who stand at the top of the food chain. Jaguars, their coats patterned with the dappled light filtering through the leaves, stalk silently through the undergrowth, their powerful bodies built for stealth and speed. They are the ultimate regulators, controlling the populations of herbivores and ensuring a delicate balance within the ecosystem. The jaguar's roar, echoing through the dense foliage, is a reminder of the power of nature's design. .

But the role of predators extends beyond mere control. They are also vital catalysts, driving evolutionary change. Their presence forces their prey to evolve, developing adaptations to evade their deadly attacks. This continuous dance of predator and prey shapes the very fabric of the rainforest, fueling its remarkable resilience. .

Beyond the mighty jaguar, a cast of smaller hunters populate the Amazonian stage. Pumas, their sleek bodies blending seamlessly with the shadows, prowl through the canopy, while ocelots, with their distinctive spotted fur, hunt in the undergrowth. Each predator occupies a specific niche, contributing to the intricate tapestry of predator-prey interactions. .

But the Amazon's mammalian cast is not solely comprised of hunters. The herbivores, with their insatiable appetites, are equally vital to the rainforest's well-being. Giant tapirs, their bodies resembling miniature rhinoceroses, lumber through the forest floor, their large snouts uprooting plants and spreading seeds. The mighty arapaima, the Amazon's freshwater leviathan, grazes on water lilies, their massive bodies stirring the murky depths. .

These herbivores are the architects of the rainforest's landscape, shaping the composition of the forest floor and contributing to the continuous cycle of growth and decay. Their grazing habits influence plant diversity and create opportunities for new species to emerge. .

The Amazon's mammalian cast is not solely defined by their dietary habits. Some, like the capybaras, the world's largest rodents, exhibit social behavior that mirrors the complex dynamics of human societies. Their large groups, living in close proximity, rely on intricate social structures to ensure the survival of their young and the efficiency of their foraging. .

Others, like the howler monkeys, utilize their vocal prowess to create a symphony of communication. Their resonant howls, echoing through the trees, serve as a territorial declaration, a means of attracting mates, and a way to coordinate within their troops. .

The nocturnal world of the Amazon is a realm of shadows and whispers, a hidden stage for creatures adapted to a life cloaked in darkness. Bats, with their sonar navigation, navigate the tangled undergrowth, their wings a blur against the moonlight. They are crucial pollinators, their tiny bodies flitting from flower to flower, ensuring the reproduction of countless plant species.

The Amazon is not only a refuge for these aerial acrobats; it also harbors a diversity of nocturnal mammals, each playing a unique role in the intricate balance of the ecosystem. The elusive olinguito, a small nocturnal carnivore, hunts for insects, while the kinkajou, with its prehensile tail, climbs through the trees in search of fruit. .

The mammals of the Amazon, from the majestic jaguar to the elusive olinguito, are more than just residents of this vast rainforest; they are the very heart of its ecosystem. Their interactions, driven by the relentless forces of evolution and natural selection, weave a tapestry of life that is both beautiful and complex. They are the architects of the rainforest's structure, the engine of its resilience, and the testament to nature's boundless creativity. Their presence is a constant reminder that life, in all its diversity, thrives through interconnectedness and interdependence. .

Threats to Amazonian Mammals: Habitat Loss and Hunting

The Amazon, a verdant tapestry woven by nature's hand, shelters a breathtaking diversity of mammals, each playing a vital role in the intricate web of life. However, this rich biodiversity faces a formidable foe – human encroachment, manifested in the twin threats of habitat loss and hunting. These forces, driven by a relentless march of development, are rapidly eroding the very foundations upon which Amazonian mammals rely for survival. .

The relentless advance of deforestation, driven by agricultural expansion, logging, and mining, fragments the once-contiguous forests, isolating populations and disrupting the delicate balance of ecosystems. As the green lung of the planet shrinks, so too does the haven for countless species. The loss of habitat translates to a domino effect, cascading through the food web, impacting prey availability and disrupting predator-prey interactions. The intricate dance of life, choreographed over millennia, is thrown into disarray, leaving many species teetering on the precipice of extinction.

Beyond the physical fragmentation of their habitat, Amazonian mammals face a silent, but no less deadly, threat: hunting. The allure of profit, driven by the demand for bushmeat, has transformed many species into commodities. From the majestic jaguar to the elusive tapir, the insatiable appetite for wild meat is pushing these creatures towards the brink. The relentless pursuit of bushmeat, fueled by poverty and market pressures, ignores the ecological consequences, leaving behind a trail of depleted populations and disrupted ecosystems.

The impact of hunting extends far beyond the immediate mortality of individuals. The selective hunting pressure on certain species, particularly those with high market value, can alter the composition of entire communities. The removal of apex predators, such as jaguars, can unleash cascading effects, disrupting prey populations and altering the dynamics of the ecosystem. The loss of these keystone species can have far-reaching consequences, impacting the entire tapestry of life within the Amazon.

The threat of habitat loss and hunting is not a singular, isolated issue, but rather a complex web of interconnected challenges. The relentless pressure from development, coupled with the lack of sustainable practices and effective conservation measures, exacerbates the vulnerability of Amazonian mammals. The consequences of these threats are far-reaching, not just for the species themselves, but for the intricate ecosystem they inhabit and the well-being of future generations.

The magnitude of the threat demands immediate action, a call to arms for the preservation of Amazonian mammals and the vital ecosystem they represent. Conservation efforts must be multi-faceted, addressing both the root causes of habitat loss and the unsustainable practices that drive hunting. This necessitates a holistic approach, encompassing protected areas, sustainable land management, and community-based conservation initiatives.

The fate of Amazonian mammals hangs in the balance, a testament to the intricate interplay between human actions and the delicate balance of nature. The responsibility lies with us, to recognize the urgent need for action, to advocate for sustainable practices, and to ensure that the symphony of life in the Amazon continues to echo through the ages. .

The urgency is palpable, the need for action undeniable. The time for complacency is over, the time for concerted effort has arrived. The fate of Amazonian mammals, and indeed the entire Amazon ecosystem, rests upon our collective will to protect this irreplaceable treasure. .

Chapter 9: The Reptiles and Amphibians: Masters of Camouflage and Survival

The Amazon's Reptile and Amphibian Diversity: A World of Camouflage and Poison

The Amazon's Reptile and Amphibian Diversity: A World of Camouflage and Poison.

The Amazon, a sprawling tapestry of verdant life, pulsates with an undercurrent of silent, scaled, and slick creatures. Its sprawling rivers, sun-drenched canopies, and damp, humid forest floor serve as a canvas for an astounding diversity of reptiles and amphibians, each a masterpiece of adaptation and survival. This vibrant world, where life teems with both beauty and danger, holds secrets whispered through the rustle of leaves and the splash of a submerged caiman. .

The Amazon's reptiles, masters of camouflage and cunning, are a testament to the power of blending in. From the emerald-hued, sun-basking iguanas to the sand-colored, sand-swimming caiman, these creatures have honed their appearance to become nearly invisible, their scales mimicking the intricate patterns of their environment. Their camouflage is more than just a visual trick; it is a survival strategy honed over millennia, allowing them to lie in wait for prey, escape predators, and navigate their world with an almost supernatural stealth. .

The Amazon's snakes, a symphony of scales and venom, weave a tale of both fear and fascination. The Anaconda, a leviathan of the water, glides through the murky depths, its massive body a formidable weapon. The Boa Constrictor, a master of stealth and pressure, wraps its powerful coils around its prey, squeezing the life out with the calculated precision of a seasoned hunter. The venomous snakes, however, play a different game, their fangs tipped with deadly toxins that paralyze and subdue their victims. The Bushmaster, a master of disguise, blends seamlessly into the undergrowth, its bite a whispered warning to those who dare cross its path. The Fer-de-Lance, a serpent of the shadows, strikes with lightning speed, its venom a potent cocktail of pain and paralysis.

The Amazon's amphibians, masters of transformation and adaptation, are a testament to the resilience of life in the face of adversity. The tree frogs, miniature acrobats of the canopy, their vibrant colors a testament to their toxicity, perch on leaves, their sticky toe pads holding them fast against the pull of gravity. The Poison Dart Frogs, tiny but deadly, their skin a tapestry of brilliant hues, are a warning to any predator that dares to approach. Their venom, a potent cocktail of alkaloids, paralyzes and kills, a potent defense against the ever-present threat of the jungle. .

The Amazon's amphibians, however, are not just masters of survival, they are also pioneers of transformation. The tadpoles, small and vulnerable, are a testament to the cycle of life, their metamorphosis a journey from aquatic to terrestrial existence. From the humble tadpole to the magnificent Amazonian Bullfrog, their transformation is a mesmerizing spectacle, a testament to the wonders of nature.

The Amazon's reptiles and amphibians are not just inhabitants of this magnificent ecosystem, they are its guardians. They control populations, maintain the delicate balance of the food web, and are crucial components of the jungle's intricate web of life. Their presence, a silent symphony of camouflage and survival, reminds us of the enduring power of adaptation, the beauty of diversity, and the fragility of this wondrous ecosystem. .

Adaptations for Survival: Camouflage, Venom, and Specialized Defenses

The Amazon, a verdant tapestry of life, teems with an astonishing array of creatures, each meticulously crafted by evolution to navigate the intricate dance of survival. Among these masters of adaptation are the reptiles and amphibians, whose strategies for survival are as diverse as the jungle itself. Camouflage, a silent art, allows them to disappear into the vibrant backdrop, blending seamlessly with their surroundings. Venom, a deadly cocktail, serves as a potent weapon, deterring predators and securing prey. Specialized defenses, a testament to nature's ingenuity, offer unique advantages in the relentless struggle for existence. .

Reptiles, with their scaly armor and ancient lineage, are masters of blending into the environment. The emerald green of the Green Anaconda (Eunectes murinus), the largest snake in the Americas, perfectly mirrors the dense foliage, allowing it to ambush unsuspecting prey. The Amazon Basin Caiman (Caiman crocodilus), with its mottled brown and green scales, practically vanishes in the murky waters of the flooded forests. The Amazon Basin Boa Constrictor (Boa constrictor), another master of camouflage, blends seamlessly into the tangled undergrowth, its patterns mimicking the dappled sunlight filtering through the canopy. .

Amphibians, with their smooth, often brightly colored skin, embrace a different approach. The Red-Eyed Tree Frog (Agalychnis callidryas), with its emerald green body and piercing red eyes, showcases disruptive coloration, breaking up its silhouette and making it appear larger than it actually is. The Amazon Milk Frog (Trachycephalus resinifictrix), with its warty skin and vibrant green and orange hues, stands out boldly, but its bright coloration serves as a warning signal to potential predators, indicating its toxicity. The Glass Frog (Centrolenidae family), with its translucent skin, almost completely disappears against the backdrop of the rainforest leaves, showcasing transparency as a powerful camouflage strategy. .

Venom, a formidable weapon forged by evolution, plays a crucial role in the survival of many Amazonian reptiles and amphibians. The Bushmaster (Lachesis muta), the largest venomous snake in the Americas, possesses a potent neurotoxin that paralyzes its prey, ensuring a swift and efficient kill. The Fer-de-Lance (Bothrops atrox), known for its aggressive nature, utilizes a hemotoxic venom that causes intense pain and tissue damage, dissuading even the most determined predator. The Amazon Basin Coral Snake (Micrurus corallinus), adorned with striking red, yellow, and black bands, is a visual testament to its deadly venom. .

Amphibians, too, have evolved venomous strategies. The Golden Dart Frog (Phyllobates terribilis), with its brilliant yellow skin, is one of the most poisonous animals on Earth. Its skin secretes a potent alkaloid toxin, Batrachotoxin, capable of paralyzing and killing even large predators. The Amazon Basin Poison Dart Frog (Dendrobates auratus), another brightly colored member of the

Dendrobatidae family, utilizes its skin toxins to deter predators, showcasing the effectiveness of chemical defense. .

Beyond camouflage and venom, reptiles and amphibians have developed a range of specialized defenses that enhance their chances of survival. The Amazon Basin Caiman Lizard (Dracaena guianensis), with its powerful tail and spiky scales, uses its formidable defenses to ward off predators, even daring to confront much larger animals. The Amazon Basin Iguana (Iguana iguana), with its sharp claws and powerful bite, can deliver a painful defensive strike, while its ability to climb trees offers an escape route from predators. The Amazon River Turtle (Podocnemis expansa), with its hard shell, provides a natural armor, offering protection against predation, while its ability to dive into the water allows it to escape potential threats. .

The Amazon Basin Toad (Bufo marinus), with its toxic skin secretions, relies on chemical defense to deter predators. The Amazon Basin Salamander (Plethodontidae family), with its ability to regenerate lost limbs, demonstrates an astounding adaptation for survival. The Amazon Basin Tree Frog (Hylidae family), with its suction cup-like toes, is able to cling tightly to vertical surfaces, offering a unique advantage in a world filled with predators. .

The intricate web of survival in the Amazonian rainforest is woven with the threads of adaptation. Reptiles and amphibians, as masters of camouflage, venom, and specialized defenses, stand as testament to the power of evolution, showcasing the remarkable diversity and ingenuity of nature's creations. Their strategies for survival, honed over millions of years, ensure their continued presence in this vibrant ecosystem, playing vital roles in the delicate balance of this extraordinary world. .

The Reptiles and Amphibians' Role in the Ecosystem: Predators, Prey, and More

The Amazon, a tapestry of vibrant life, is woven with threads of intricate relationships between all its inhabitants. Amidst the towering trees and the swirling waters, reptiles and amphibians play a vital role, shaping the ecosystem in ways often unseen but undeniably crucial. Their lives, marked by adaptation and survival, are intricately intertwined with the fate of countless other species.

Reptiles, with their ancient lineage and diverse forms, are masters of camouflage, seamlessly blending into their environment. The emerald green boa constrictor, a phantom in the undergrowth, patiently awaits unsuspecting prey. Its skin, a living mosaic of emerald and jade, mimics the dappled light filtering through the canopy, making it virtually invisible to the untrained eye. This mastery of camouflage allows the boa to ambush its prey, effectively transforming it into a predator lurking in the shadows.

Amphibians, the denizens of both water and land, have honed a different strategy for survival. The poison dart frog, a jewel of vibrant colors, uses its bright hues as a warning sign, signaling its toxic nature to potential predators. Its vibrant red, yellow, and blue markings are a stark reminder of its potent defense mechanism, effectively deterring most predators from attempting to consume it. This bold display of color, rather than serving as a camouflage, is a powerful tool for survival, ensuring the frog's safety in the competitive jungle environment.

The role of reptiles and amphibians extends beyond their individual survival strategies. They are, in many cases, the crucial link in the delicate balance of the Amazonian food web. The caiman, a formidable predator, patrols the murky waters, its presence acting as a natural regulator of fish populations. It controls the numbers of certain fish species, preventing overgrazing of aquatic plants and maintaining a healthy balance in the aquatic ecosystem. This top-down regulation is essential for the health and stability of the entire Amazonian web of life.

Amphibians, with their diverse dietary habits, play a vital role in controlling insect populations. The Amazonian tree frog, with its sticky tongue and agile movements, is a voracious predator of insects, keeping their numbers in check. By controlling insect populations, amphibians help prevent outbreaks of pests and maintain the delicate balance of the rainforest ecosystem. .

Beyond their roles as predators and prey, reptiles and amphibians contribute to the nutrient cycle in the Amazon. The venomous coral snake, a striking example of natural beauty, is a crucial link in the decomposition process. Its venom, while deadly to prey, breaks down the carcass, releasing vital nutrients back into the ecosystem, enriching the soil and nourishing the rainforest's flora.

The Amazon, a vast and intricate tapestry of life, relies on the diverse roles of its inhabitants. Reptiles and amphibians, with their unique adaptations and survival strategies, contribute significantly to this complex ecosystem. From their role as apex predators to their intricate contributions to nutrient cycling, these creatures are essential for the health and stability of the Amazon rainforest. Their survival, intimately linked to the survival of countless other species, is a testament to the interconnected nature of life in the Amazon.

Threats to Amazonian Reptiles and Amphibians: Habitat Loss and Climate Change

The Amazon rainforest, a sprawling tapestry of life, pulsates with a diversity that captivates the imagination. Within its emerald embrace, a silent drama unfolds: the battle for survival of its reptilian and amphibian inhabitants. These masters of camouflage, these living embodiments of ancient lineages, find themselves increasingly besieged by the encroaching tide of human activity. The very fabric of their existence, the rainforest itself, is unraveling under the relentless assault of habitat loss and climate change.

The Amazon, with its tangled undergrowth and shimmering waterways, is a sanctuary for an extraordinary array of reptiles and amphibians. The vibrant emerald green of the emerald boa, its skin a testament to the rainforest's vibrancy, blends seamlessly with the foliage as it hunts its prey. The iridescent blue of the poison dart frog, a walking chemical factory, serves as a stark warning to predators, a testament to the intricate interplay of evolution and defense. But beneath the surface of this vibrant spectacle, a quiet crisis is unfolding.

The specter of habitat loss looms large over the Amazonian reptilian and amphibian populations. The relentless expansion of agriculture, logging, and mining activities is steadily carving out swathes of the rainforest, fracturing the delicate balance of this ecosystem. The loss of their natural habitat translates to a loss of food sources, breeding grounds, and shelter. The consequences are stark: dwindling populations, fragmented ranges, and a heightened vulnerability to extinction.

The Amazon, a region that thrives on a delicate balance of water cycles and humidity, is facing the brunt of climate change. Rising temperatures, altered rainfall patterns, and increased frequency of extreme weather events are disrupting the ecosystem, pushing reptiles and amphibians to the brink. The delicate dance of reproduction, dependent on specific environmental cues, is thrown into disarray. Nesting sites are destroyed, breeding seasons disrupted, and the very air they breathe becomes a hostile environment.

The impact of climate change is compounded by the inherent vulnerability of reptiles and amphibians. Their ectothermic nature, their reliance on external sources of heat, makes them particularly susceptible to temperature fluctuations. As temperatures rise, their physiological functions are disrupted, leading to decreased activity, slowed growth, and increased vulnerability to diseases. The delicate balance of their skin, often permeable and crucial for respiration and water absorption, is further compromised by changing climatic conditions, making them susceptible to environmental toxins and dehydration.

The fate of the Amazon's reptilian and amphibian inhabitants is intertwined with the fate of the rainforest itself. Their survival hinges on the preservation of their habitat, the very essence of their existence. The ongoing deforestation, driven by unsustainable agricultural practices, logging, and mining, is a stark reminder of the human impact on this fragile ecosystem. The loss of habitat not only deprives them of vital resources but also forces them to adapt to fragmented landscapes, further increasing their vulnerability to predation, competition, and disease.

The threat of climate change further complicates the equation. The Amazon, a vital carbon sink, is rapidly losing its ability to absorb atmospheric carbon dioxide due to deforestation and rising temperatures. This loss of carbon sequestration exacerbates climate change, creating a vicious cycle that further threatens the rainforest and its inhabitants. The rising temperatures, altered rainfall patterns, and increased frequency of extreme weather events are altering the delicate balance of this ecosystem, pushing reptiles and amphibians to the brink.

The story of the Amazonian reptiles and amphibians is not just a tale of ecological loss, but a stark reminder of the interconnectedness of life on Earth. The consequences of habitat loss and climate change ripple through the entire ecosystem, impacting not only the individual species but also the intricate web of relationships that sustain the rainforest's biodiversity. .

As the Amazon continues to be transformed, the fate of its reptilian and amphibian inhabitants hangs in the balance. Their survival depends on our collective commitment to sustainable practices, a shift towards environmental responsibility, and a recognition of the critical role they play in the delicate tapestry of life. The future of the Amazon, and the creatures that call it home, rests on our ability to act now, to ensure that the symphony of life continues to play its intricate melody. .

Chapter 10: The Insects: A Buzzing World of Life

The Amazon's Insect Diversity: A Microcosm of Life

The Amazon's Insect Diversity: A Microcosm of Life.

The Amazon rainforest, a sprawling tapestry of emerald green woven through with the vibrant threads of life, is not merely a spectacle of towering trees and exotic animals. It is also a teeming microcosm of life, a vibrant ecosystem that teems with an astonishing diversity of insects. The sheer scale of this insect life is mind-boggling, a staggering testament to nature's ability to populate even the most extreme environments. Imagine a world where the buzzing of countless wings fills the air, a symphony of life that echoes through the dense canopy and across the murky waters. This is the Amazon, a place where insect diversity is not merely a curiosity, but a fundamental pillar of the ecosystem's intricate web.

The Amazon, with its vast expanse, serves as a cradle for an astonishing array of insect species. Scientists estimate that millions of insect species call the Amazon home, a figure dwarfed only by the equally vast and largely unexplored world of microorganisms. The very diversity of the insects themselves reflects the incredible ecological richness of the rainforest. From the minuscule ants that tunnel through the forest floor to the imposing beetles that adorn the rainforest's canopies, each species plays a unique role, intricately connected to the delicate balance of the ecosystem.

The Amazon's insect diversity is not simply a random collection of species. It is a complex web of relationships, a tapestry woven with threads of symbiosis, predation, and competition. The very existence of these diverse species depends upon a intricate balance of these relationships, a delicate dance that has been shaped over millions of years. For instance, the vibrant colors of some butterfly species serve as a warning to potential predators, a signal of their toxic or unpleasant taste. This adaptation, a product of natural selection, demonstrates the crucial interplay between predators and prey, a dynamic force shaping the very evolution of insect diversity in the Amazon.

Beyond the mesmerizing dance of evolution, the sheer number of insects in the Amazon plays a vital role in maintaining the rainforest's ecosystem. These tiny creatures are the unsung heroes of the jungle, tirelessly performing essential tasks that keep the delicate balance of life intact. Imagine a vast army of decomposers, tirelessly breaking down fallen leaves and deadwood, returning nutrients to the soil that nourishes the forest's giants. This is the vital work of insects, enriching the soil and fueling the cycle of life in the Amazon.

Furthermore, insects are crucial pollinators, ensuring the reproduction of countless plant species. As they flit from flower to flower, these tiny ambassadors of life carry pollen grains, enabling the fertilization of plants and the perpetuation of the rainforest's rich floral tapestry. The very existence of the Amazon's diverse flora depends heavily on the tireless work of these tiny pollinators, a testament to the interconnectedness of life within the rainforest.

However, this intricate web of life, so delicately balanced, is facing increasing threats. The Amazon's insect diversity, once a vibrant tapestry of life, is now under threat from the relentless march of deforestation and habitat destruction. As the forest's canopy is cleared, the very fabric of the ecosystem unravels, leaving the insect world vulnerable to the cascading effects of habitat loss and fragmentation. This not only affects the insects themselves but also disrupts the intricate web of life, impacting the entire ecosystem.

Beyond the loss of habitat, the Amazon's insect populations are also under threat from the insidious encroachment of agricultural practices. The widespread use of pesticides, designed to control pests in agricultural fields, often spills over into the surrounding rainforest, impacting the insect populations that play a vital role in maintaining the ecosystem's delicate balance. This indiscriminate use of chemicals disrupts the food chains, endangering not only the target pests but also the countless beneficial insects that contribute to the rainforest's intricate web of life.

The Amazon's insect diversity, a marvel of nature, is facing unprecedented challenges, a stark reminder of the fragility of life within the rainforest. As the world grapples with the consequences of human activities, it is crucial to understand and appreciate the vital role insects play in maintaining the delicate balance of the Amazon's ecosystem. Preserving this intricate web of life is not just a matter of scientific curiosity; it is an imperative for the future of the Amazon and the planet as a whole. .

The Insects' Role in the Ecosystem: Pollination, Decomposition, and Food Chains

The Insects' Role in the Ecosystem: Pollination, Decomposition, and Food Chains.

The Amazon rainforest, a tapestry of vibrant life, is a testament to the intricate interplay of countless species. Among these, the insects hold a position of paramount importance, their tiny bodies orchestrating vital processes that underpin the entire ecosystem. Their role extends far beyond mere buzzing and crawling, influencing everything from the intricate dance of pollination to the silent recycling of organic matter.

Imagine a world without vibrant flowers, their colors and fragrances muted, their pollen forever trapped within their petals. This is the reality that would face the Amazon without its insect pollinators. Bees, butterflies, moths, and even some flies diligently move from flower to flower, their bodies dusted with pollen, transferring the precious genetic material that allows plants to reproduce. This intricate dance of pollination ensures the continued existence of countless plant species, enriching the biodiversity of the rainforest and providing sustenance for herbivores and ultimately, for the entire food chain. .

The Amazon is a symphony of decay, with fallen leaves, dead animals, and decaying wood constantly providing a rich source of nutrients. This is where the insects step in as nature's recyclers. Beetles, termites, ants, and a myriad of other insects break down the organic matter, transforming it into soil-enriching humus. This decomposition process releases vital nutrients back into the ecosystem, fueling the growth of new life and creating the fertile ground upon which the rainforest thrives. Without these tireless scavengers, the Amazon would be choked with decaying matter, its nutrient cycle disrupted, and its ability to sustain life severely compromised.

The role of insects in the Amazonian food chain is multifaceted and complex, intertwining predator and prey in a delicate balance. These tiny creatures form the foundation of the food chain, providing sustenance to countless birds, reptiles, amphibians, and mammals. A single caterpillar, munching on a leaf, may become the prey of a spider, which in turn might be consumed by a dart frog. This cascade of energy transfer ensures the survival of countless species, demonstrating the profound influence of insects on the Amazonian ecosystem. .

Furthermore, the insects themselves are preyed upon by a diverse array of predators, contributing to a complex and balanced food web. From the humble praying mantis, camouflaged amongst the leaves, to the swift swooping of a kingfisher, the hunt for insects provides a crucial source of energy for a multitude of rainforest inhabitants. This intricate interplay between predator and prey ensures the survival of both sides, fostering a healthy and resilient ecosystem.

The Amazon, a treasure trove of biodiversity, owes its continued existence to the tireless work of its insect inhabitants. From the vibrant ballet of pollination to the silent decomposition of organic matter, insects weave a tapestry of life, ensuring the rainforest remains a vibrant and resilient ecosystem. As we explore the Amazon, we must acknowledge the vital role played by these often overlooked creatures, for without them, the rainforest would cease to be the vibrant and magnificent place it is today. .

The Importance of Insects: A Keystone to Life in the Jungle

The Amazon jungle, a verdant tapestry of life, thrives on a hidden symphony of buzzing, crawling, and flitting creatures. The insects, often overlooked in the grand spectacle of the rainforest, are not merely a backdrop but the very cornerstone of this ecosystem. They are the architects of its intricate web of life, the unseen hands that weave together the delicate threads of survival and propagation. Their sheer abundance, diversity, and interconnectedness make them an indispensable force in maintaining the Amazon's delicate balance, a testament to the power of small, often-unnoticed beings. .

Their role begins with the very foundations of the rainforest. From the towering canopy to the muddy riverbanks, insects are the tireless decomposers, breaking down fallen leaves, decaying wood, and animal carcasses. This constant process of decomposition, fueled by the tireless efforts of beetles, ants, termites, and countless other species, releases vital nutrients back into the soil, nourishing the vibrant jungle growth. Without their diligent work, the rainforest would be choked by its own debris, a cycle of life abruptly halted.

The intricate tapestry of the jungle food web is woven through the actions of insects. They are a vital food source for countless species, from the tiny dart frog to the majestic jaguar. Birds feast on brightly colored butterflies, while spiders build intricate webs to ensnare unsuspecting flies and moths. The Amazon's diversity of life, from the smallest frog to the largest monkey, is intricately tied to the abundance and variety of insects. Their presence provides a foundation for the entire ecosystem, ensuring the survival of countless species.

Insects are also the unseen drivers of plant reproduction. Pollination, the transfer of pollen between flowers, is a crucial step in the lifecycle of most flowering plants. Bees, butterflies, moths, and even some flies are the tireless messengers of this process, transferring pollen from flower to flower, ensuring the production of seeds and fruits. The intricate dance of pollination, driven by the tireless work of insects, ensures the continuation of the rainforest's vibrant tapestry of plant life. .

The impact of insects extends far beyond the immediate ecosystem. They are the silent architects of soil fertility, aerating the earth with their tunnels and enriching it with their waste products. This constant process of soil rejuvenation ensures the continued growth of the rainforest, providing the foundation for all life within it. .

The complex interactions between insects and other organisms create a delicate balance that ensures the health and resilience of the Amazon. For example, the presence of insect predators, like spiders and dragonflies, keeps the populations of herbivorous insects in check, preventing them from decimating the delicate rainforest vegetation. This natural regulation ensures the sustainability of the ecosystem, preventing any single species from dominating the intricate web of life. .

The Amazon's insects are not merely passive participants in this grand spectacle of life; they are active players, shaping the very landscape and driving the intricate processes that sustain the rainforest. Their intricate relationships with plants, animals, and the soil itself demonstrate the power of seemingly insignificant creatures to influence the fate of an entire ecosystem. They are the silent architects of the Amazon's biodiversity, their buzzing, crawling, and flitting a symphony of life that resonates through the very heart of the rainforest. .

The vast diversity of insect life in the Amazon, with its thousands of species, reflects the complexity and richness of the rainforest itself. Each insect, from the smallest gnat to the largest beetle, plays a unique role in the intricate tapestry of life, contributing to the overall health and resilience of the ecosystem. The Amazon's insects are a testament to the power of small creatures, a reminder that even the smallest beings can shape the fate of an entire world. .

From the towering canopy to the muddy riverbanks, the Amazon's insects are an indispensable force. Their vital contributions to pollination, decomposition, and the food web underscore their crucial role in maintaining the delicate balance of the rainforest. Their sheer abundance and diversity serve as a constant reminder of the intricate interconnectedness of life, a testament to the power of nature's delicate balance. As we navigate the Amazon's vibrant tapestry of life, it is essential to acknowledge the crucial contributions of these often-overlooked creatures, for they are the unsung heroes of the jungle, the buzzing, crawling, and flitting architects of a world teeming with life. .

Threats to Amazonian Insects: Habitat Loss and Pesticides

The Amazon rainforest, a tapestry of life woven into a vibrant emerald expanse, teems with a staggering diversity of insects. These miniature marvels, often overlooked in the grander narratives of the jungle, play vital roles in the intricate web of life. From the tireless pollinators to the soil-enriching decomposers, they are the very foundation of the rainforest's ecological stability. However, this buzzing world of life faces a chilling reality - a reality defined by the encroaching threats of habitat loss and pesticide use, both driven by human actions.

The relentless advance of deforestation, fueled by agricultural expansion, logging, and mining operations, is carving deep wounds into the very fabric of the rainforest. The loss of habitat is akin to pulling the rug out from under the feet of these insects, disrupting their delicate existence. The intricate tapestry of life that once thrived within these forests is unraveling, leaving behind a fragmented landscape with limited resources.

The impact on insect populations is profound. As their homes are destroyed, insects lose access to vital food sources, nesting grounds, and shelter. The once-thriving populations are forced to fragment, shrink, and in some cases, disappear entirely. This fragmentation not only weakens individual populations but also disrupts their complex interactions within the ecosystem.

The intricate dance of pollination, a cornerstone of the rainforest's existence, is thrown into disarray. Many plants rely on specific insects for pollination, and the loss of these crucial pollinators can trigger a cascade of consequences. The delicate balance between plants and insects, a relationship that has evolved over millennia, is strained to its breaking point.

Beyond the immediate threat of habitat loss, the insidious presence of pesticides further complicates the situation. The overuse and misuse of pesticides, a common practice in agricultural areas encroaching upon the rainforest, act as silent assassins, decimating insect populations. The chemicals, designed to target specific pests, often kill non-target insects, disrupting the delicate equilibrium of the ecosystem.

The ramifications of pesticide use are far-reaching. The chemicals can contaminate water sources, impacting the aquatic life that supports a myriad of insect species. They can also accumulate in the food chain, affecting predator populations that rely on insects as their food source. The consequences extend beyond the immediate mortality of insects; they ripple through the entire ecosystem, potentially leading to cascading failures.

The combined pressure of habitat loss and pesticide use is pushing insect populations towards a precipice. The delicate balance of the Amazonian ecosystem is being tipped, with far-reaching consequences for the rainforest's biodiversity and the very foundation of its existence. As we witness the alarming decline of these miniature marvels, we must recognize the profound impact they have on the rainforest's intricate web of life. It is imperative that we prioritize conservation efforts, fostering sustainable practices that protect both the habitats and the insects that call them home. The fate of the Amazonian insects, and ultimately, the fate of the entire rainforest, hangs in the balance. .

Chapter 11: The Fungi: Masters of Decomposition

The Amazon's Fungi Diversity: A Hidden Realm of Life

The Amazon's Fungi Diversity: A Hidden Realm of Life.

The Amazon rainforest, a vast and verdant tapestry woven across the South American continent, is renowned for its unparalleled biodiversity. However, beneath the canopy, hidden from casual observation, lies a realm of life as diverse and intricate as the towering trees above. It is the realm of fungi, a kingdom of microscopic marvels playing an indispensable role in the intricate web of life that sustains this ecosystem.

The Amazon's fungal diversity is staggering. Estimates suggest millions of fungal species reside within this ecosystem, yet only a fraction have been scientifically documented. This hidden kingdom thrives in every nook and cranny of the rainforest, from the damp forest floor to the decaying wood of fallen giants. Their presence is not confined to the shadows, however, as some species form intricate symbiotic relationships with plants, forming mycorrhizal networks that extend far beyond the visible realm.

Fungi are the masters of decomposition, the silent architects of nutrient recycling within the Amazon. They break down dead organic matter, releasing vital nutrients back into the soil, fueling the growth of plants and sustaining the intricate food web of the rainforest. This role is critical to the Amazon's remarkable productivity, allowing the ecosystem to thrive despite the relentless cycle of decomposition and regeneration.

The sheer diversity of Amazonian fungi reflects the multitude of ecological niches they occupy. From the microscopic, single-celled yeasts fermenting fruit and leaves, to the massive, intricate fruiting bodies of giant mushrooms, each species has evolved unique adaptations to thrive in the specific microenvironment it inhabits. This diversity reflects the complexity and resilience of the rainforest ecosystem, showcasing the intricate interplay of life forms and their dependence on one another.

One of the most striking aspects of Amazonian fungi is their intricate relationship with plants. Mycorrhizal fungi form symbiotic relationships with plant roots, extending their reach into the soil and facilitating nutrient uptake. In return, the fungi receive sugars from the plant, creating a mutually beneficial exchange that sustains both partners. This relationship is vital for the survival of many rainforest plants, particularly in the nutrient-poor soils characteristic of the Amazon.

The diversity of mycorrhizal fungi in the Amazon is staggering, with each plant species often harboring multiple fungal partners. This diversity reflects the complex and dynamic nature of these symbiotic relationships, showcasing the intricate web of connections that underpin the Amazon's ecological stability. .

Beyond their ecological role, Amazonian fungi hold immense cultural significance for indigenous communities. Many species are used in traditional medicines, while others serve as sources of food and dyes. The intricate knowledge of fungi possessed by these communities underscores the deep connection between humans and the natural world, highlighting the importance of preserving both cultural and biological diversity.

The study of Amazonian fungi is still in its infancy. Every foray into the rainforest unveils new species, revealing the vastness and complexity of this hidden world. Scientists are uncovering the potential of fungi in diverse fields, from bioremediation to biofuel production, further emphasizing the importance of understanding this fascinating kingdom.

The Amazon's fungi are not just a hidden realm; they are a vital component of the rainforest's intricate web of life. Their presence underscores the interconnectedness of all living things, reminding us that the seemingly insignificant can hold immense power and potential. As we delve deeper into the mysteries of this hidden kingdom, we gain a richer understanding of the Amazon's resilience, complexity, and the vast potential it holds for future discoveries. .

The Fungi's Role in the Ecosystem: Breaking Down Dead Matter and Recycling Nutrients

The Amazon rainforest, a symphony of life teeming with vibrant flora and fauna, is not just about the dazzling spectacle above ground. Beneath the leafy canopy, hidden from the casual observer, lies a silent world of unseen activity. This world, dominated by the fungi, is an unseen powerhouse, responsible for the ceaseless recycling of nutrients that fuels the rainforest's magnificent ecosystem. .

While plants are the primary producers, converting sunlight into energy, fungi are the essential recyclers, breaking down dead organic matter and returning vital nutrients back to the soil. Their role in this complex web of life is absolutely critical. Without fungi, the rainforest would be choked with decaying organic matter, preventing the growth of new life. .

The sheer diversity of fungal species in the Amazon is astounding, with estimates suggesting millions of undiscovered species lurking beneath the forest floor. This incredible diversity reflects their wide range of adaptations, each species uniquely tailored to break down specific types of organic matter. Some specialize in decomposing wood, while others tackle leaves, fruits, and even animal carcasses. .

Fungi employ a remarkable arsenal of enzymes to break down the tough, complex molecules found in dead organic matter. These enzymes, secreted into the surrounding environment, act like molecular scissors, dismantling the complex structures of lignin, cellulose, and other recalcitrant compounds. The breakdown products, simpler sugars and nutrients, are then absorbed by the fungi, providing them with energy and building blocks for growth.

This process of decomposition isn't just about clearing out dead matter. It is the very foundation of nutrient cycling, a critical process that sustains the rainforest's astonishing biodiversity. As fungi break down dead organisms, they release vital nutrients like nitrogen, phosphorus, and potassium back into the soil. These nutrients, essential for plant growth, are then absorbed by the roots of trees and other plants, ensuring the continued flourishing of the forest. .

The intricate dance of decomposition and nutrient recycling is a delicate balance, essential for the health and stability of the rainforest ecosystem. Fungi are the orchestrators of this intricate process, playing a vital role in maintaining the flow of nutrients and ensuring the long-term sustainability of the Amazon. .

The importance of fungi in the Amazon rainforest goes far beyond their role in decomposition. They form intricate relationships with plants, both beneficial and detrimental. Mycorrhizal fungi, for example, enter into symbiotic partnerships with the roots of trees, extending their reach into the soil and providing them with access to essential nutrients. In return, the fungi receive sugars from the plants, creating a mutually beneficial relationship that is crucial for the health and growth of both partners. .

However, not all fungal relationships are so harmonious. Some fungi act as pathogens, attacking living plants and causing disease. These parasitic fungi can have devastating consequences, weakening individual trees and even causing entire populations to decline. However, even these destructive fungi play a role in the ecosystem, ultimately contributing to the complex dynamics of life and death that shape the Amazon rainforest. .

The fungal world of the Amazon is a hidden universe of incredible diversity and intricate relationships. These seemingly simple organisms are crucial to the entire ecosystem, playing a vital role in decomposition, nutrient cycling, and the balance of life and death. Studying these fascinating fungi provides invaluable insights into the intricate workings of the rainforest, highlighting its complex web of interactions and the delicate equilibrium that sustains this awe-inspiring ecosystem. .

The impact of human activities, such as deforestation and climate change, poses a significant threat to the fungal communities of the Amazon. Habitat loss and alterations to the rainforest ecosystem can disrupt the delicate balance of decomposition and nutrient cycling, potentially leading to cascading effects throughout the entire rainforest. .

Understanding the role of fungi in the Amazon is crucial for preserving this vital ecosystem. By studying and understanding these silent masters of decomposition, we can gain a deeper appreciation for the intricate web of life that sustains the rainforest and develop strategies to protect it for future generations. .

The Amazon rainforest is a treasure trove of biodiversity, a vibrant tapestry of life that extends far beyond the visible world. Beneath the surface, in the realm of decomposition and nutrient cycling, lies a hidden world dominated by fungi – the unseen masters of this complex and essential process. By delving into the fascinating world of fungi, we gain a deeper understanding of the interconnectedness of life and the delicate balance that sustains this remarkable ecosystem. .

The Importance of Fungi: Supporting Life in the Jungle

The Amazon rainforest, a tapestry woven with vibrant life, owes its extraordinary biodiversity to a silent, unseen force - fungi. These remarkable organisms, often overlooked in the grandeur of towering trees and vibrant flora, are the unsung heroes of the jungle, silently orchestrating the intricate dance of life and death. .

Fungi, far from being mere decomposers, are the architects of the Amazon's fertile soil. Their intricate network of hyphae, woven through the forest floor, acts as a living sieve, breaking down the fallen leaves, dead wood, and animal remains into their essential components. This intricate process, known as decomposition, is the lifeblood of the rainforest, releasing nutrients locked within decaying matter back into the ecosystem, fueling the growth of new life. .

Without fungi, the Amazon would be a graveyard of its own making. The forest floor would be choked with decaying matter, hindering the growth of new plants and ultimately choking out the life that thrives within its canopy. This is where the remarkable abilities of fungi truly shine. They possess a unique metabolic machinery, capable of breaking down complex organic molecules, such as lignin and cellulose, which are indigestible by most organisms. .

This remarkable feat is achieved through the production of a diverse array of enzymes. Each species of fungi has its own arsenal of enzymes, specifically tailored to break down certain types of organic matter. This intricate dance of specialized enzymes allows for the efficient decomposition of fallen logs, decaying leaves, and even animal carcasses, releasing a symphony of nutrients back into the soil. .

Beyond their role in decomposition, fungi are vital for the health of the rainforest's flora. They form symbiotic relationships with plants, known as mycorrhizae, extending their reach through intricate networks of hyphae, effectively expanding the roots' absorptive surface area. This symbiotic partnership allows plants to access nutrients, such as phosphorus and nitrogen, which are often scarce in the nutrient-poor soils of the rainforest. .

The intricate web of fungal hyphae also acts as a conduit for communication between plants, allowing them to share resources and even warn each other of impending threats. This intricate network, known as the "wood wide web," underscores the interconnectedness of the Amazon's ecosystem, highlighting the vital role fungi play in maintaining its delicate balance.

The Amazon's fungi are not just silent partners in the ecosystem; they also serve as a vital food source for countless animals. From insects and millipedes to armadillos and tapirs, a diverse array of creatures rely on fungi for sustenance. The intricate networks of fungal hyphae also provide a habitat for a multitude of microscopic organisms, further enriching the biodiversity of the rainforest floor.

The role of fungi extends beyond the tangible benefits of decomposition and nutrient cycling. They are also intricately woven into the cultural tapestry of the Amazon's indigenous communities. Many tribes utilize fungi in traditional medicine, harnessing their potent medicinal properties to treat ailments ranging from infections to skin diseases. The use of fungi in traditional practices underscores the profound connection between humans and these vital organisms, recognizing their essential role in maintaining the rainforest's delicate balance. .

The importance of fungi in the Amazon rainforest cannot be overstated. These often-overlooked organisms, silently orchestrating the complex web of life, are the architects of the rainforest's fertility, the guardians of its biodiversity, and the unseen thread that connects its intricate tapestry. They are a testament to the hidden wonders of the natural world, reminding us that even the smallest and most inconspicuous organisms play vital roles in sustaining the delicate balance of life on Earth. .

Threats to Amazonian Fungi: Habitat Loss and Pollution

Threats to Amazonian Fungi: Habitat Loss and Pollution.

The Amazon rainforest, a realm of unparalleled biodiversity, harbors a hidden world beneath its leafy canopy—a world of fungi. These enigmatic organisms, often overlooked, are the silent architects of decomposition, playing a critical role in the rainforest's intricate web of life. However, the very forces that threaten the Amazon's iconic flora and fauna also cast a shadow over these fungal communities, jeopardizing their existence and the ecological balance they maintain.

Habitat loss, driven by relentless deforestation and agricultural expansion, poses a primary threat to Amazonian fungi. The intricate mosaic of habitats, from towering trees to decaying logs, provides a diverse array of niches for fungal colonization. However, as forests are cleared for agricultural land or succumb to wildfires, these niches disappear, leading to a dramatic decline in fungal diversity. Deforestation not only removes the physical habitat but also alters the microclimate, disrupting the delicate balance of humidity, temperature, and light that fungi rely upon for growth and survival.

Beyond habitat loss, the relentless march of human activity introduces a multitude of pollutants into the Amazonian ecosystem, further jeopardizing fungal communities. Agricultural runoff laden with fertilizers and pesticides seeps into waterways, contaminating the soil and water that fungi depend on. Mining operations, a pervasive threat across the Amazon, release heavy metals and other toxic substances, poisoning the environment and disrupting fungal populations. The insidious effects of air pollution, stemming from deforestation and industrial activities, also infiltrate the rainforest, impacting fungal growth and reproduction.

The impact of these threats extends beyond individual fungal species, cascading through the rainforest ecosystem. Fungi are essential partners in symbiotic relationships with plants, forming mycorrhizal networks that facilitate nutrient uptake and enhance plant growth. As fungal diversity declines, these symbiotic relationships become strained, potentially impacting the health and resilience of Amazonian trees. The loss of fungal decomposers further disrupts the intricate nutrient cycling processes, hindering the breakdown of organic matter and the release of vital nutrients.

The consequences of fungal decline in the Amazon are far-reaching, potentially undermining the very foundation of the rainforest ecosystem. The loss of fungal diversity could disrupt plant communities, reduce carbon sequestration, and exacerbate the effects of climate change. Furthermore, the impact extends to the countless organisms that depend on fungi for sustenance, including insects, mammals, and even humans.

The future of Amazonian fungi hinges on our ability to address the underlying drivers of habitat loss and pollution. Sustainable land management practices, including reduced deforestation, responsible agricultural practices, and strict regulations on mining activities, are crucial for mitigating these threats. Conservation efforts must extend beyond protecting the iconic megafauna and embrace the often-invisible world of fungi, ensuring the preservation of their vital roles in the Amazon's intricate ecosystem.

The fate of the Amazon's fungal communities, like the fate of the rainforest itself, hangs in the balance. Our collective responsibility to safeguard this precious ecosystem demands a proactive and comprehensive approach that addresses both the visible and the hidden threats, ensuring the continued existence of these silent masters of decomposition and the vital role they play in the Amazon's vibrant tapestry of life. . .

Chapter 12: The Plants: A Symphony of Colors and Textures

The Amazon's Plant Diversity: A World of Beauty and Adaptability

The Amazon's Plant Diversity: A World of Beauty and Adaptability.

The Amazon rainforest, a vast tapestry of emerald green stretching across South America, holds within its verdant embrace a staggering array of plant life. This botanical wonderland, a testament to the power of evolution and adaptation, is a symphony of colors, textures, and forms that constantly challenges the imagination. Each leaf, flower, and tree embodies a story of survival, a silent narrative woven through millennia of interaction with the rainforest's unique and challenging environment. .

The Amazon's plant diversity is not simply a matter of quantity; it is a testament to the remarkable adaptability of life. This diversity is a result of a complex interplay of factors, including a vast and varied geographical landscape, a consistently warm and humid climate, and a rich tapestry of soil types. The Amazon's rivers, with their meandering paths and fluctuating water levels, further contribute to the diverse array of habitats that support this extraordinary plant life. .

The sheer number of species present is staggering. Estimates suggest that the Amazon holds over 40,000 plant species, representing nearly 10% of the world's total flora. This remarkable figure underscores the Amazon's position as a global hotspot for biodiversity, and it highlights the importance of preserving this precious ecosystem. Within this vast collection, we find an extraordinary spectrum of life forms, each uniquely adapted to thrive in the Amazon's demanding environment. .

The rainforest's canopy, a vibrant green roof, is home to a fascinating assortment of epiphytes, plants that live on other plants for support. These aerial acrobats, including orchids, bromeliads, and ferns, have evolved specialized mechanisms to capture water and nutrients from the air and the host plant. Their vibrant colors and intricate shapes are a testament to their success in harnessing the resources of their unusual habitat.

Below the canopy, the forest floor is a realm of dappled light and decaying vegetation. Here, a diverse array of plants, including ferns, mosses, and lichens, thrive in the damp, nutrient-rich environment. These plants, often overlooked in the dazzling display of the canopy, play a critical role in the rainforest ecosystem, contributing to nutrient cycling and providing habitat for a host of small creatures.

The Amazon's rivers and floodplains are home to another unique group of plants: the aquatic and semi-aquatic species. These resilient plants have adapted to the fluctuating water levels of the rivers, developing specialized roots and stems to cope with periods of submersion and exposure. Water lilies, with their large, floating leaves and vibrant blossoms, are a familiar sight in the Amazon's waterways, while the towering palm trees that fringe the riverbanks provide vital habitat for a variety of wildlife.

The Amazon's plant life is not merely a collection of beautiful and unusual species; it is a complex web of interrelationships, each species playing a vital role in the rainforest's delicate balance. From the towering trees that provide shade and structure to the tiny, often overlooked plants that contribute to the soil's fertility, every species is interconnected, contributing to the overall health and resilience of this remarkable ecosystem. .

The Amazon's plants are also a vital source of sustenance for the indigenous communities that call this region home. For centuries, these people have relied on the rainforest's bounty, using its plants for food, medicine, and building materials. The Amazon's flora represents a rich cultural heritage, woven into the fabric of indigenous life.

The diversity of the Amazon's plant life extends beyond the rainforest itself. The surrounding savannas and grasslands, with their unique environmental conditions, support a distinct collection of plants. These species, often adapted to drought and fire, play a vital role in the region's ecosystem, providing food and habitat for a variety of animals and contributing to the overall ecological balance.

As we delve deeper into the Amazon's plant diversity, we discover a world of wonder and complexity, a realm where beauty and function are intertwined in a harmonious symphony of life. Every leaf, every flower, and every tree is a testament to the remarkable adaptability of life, a reminder of the importance of preserving this precious and irreplaceable ecosystem for future generations. .

Adaptations for Survival in the Jungle: From Photosynthesis to Defense Mechanisms

The Amazon rainforest, a verdant tapestry woven with life, is a testament to the boundless creativity of nature. The sheer diversity of plant life within this ecosystem is a source of endless fascination for the naturalist, each species a masterpiece of adaptation honed over millions of years to thrive in the demanding conditions of the jungle. This symphony of colors and textures, a vibrant orchestration of form and function, reveals a story of survival, each element meticulously crafted to secure a foothold in this competitive and dynamic environment. .

From the towering canopy giants to the diminutive herbs clinging to the forest floor, the Amazon's plant life exhibits an astonishing array of strategies for survival, each one a masterpiece of natural engineering. The struggle for sunlight, a precious resource in the dense understory, has sculpted leaves into an array of shapes and sizes, maximizing exposure to the precious rays that pierce through the canopy. The broad, flat leaves of the giant water lily, Victoria amazonica, provide a vast surface for absorbing sunlight, while the slender, pointed leaves of the rainforest understory plants, such as the ferns and mosses, navigate the dappled light, capitalizing on every available ray. .

The struggle for water, another vital resource in the rainforest's capricious climate, has shaped roots into complex networks that tap into the shallow topsoil and even extend beyond the reach of the dense canopy. The shallow root systems of epiphytes, like the orchids and bromeliads, cling to tree trunks and branches, collecting moisture from rain and humidity, while the deep roots of trees like the Brazil nut tree, Bertholletia excelsa, anchor themselves firmly to the soil, drawing sustenance from the rich earth.

The challenge of nutrients, limited by the rapid decomposition of organic matter in the hot, humid climate, has led to fascinating strategies. Plants like the nitrogen-fixing legumes, such as the Amazonian wild bean, Inga edulis, harbor symbiotic bacteria in their roots, converting atmospheric nitrogen into usable forms. This allows them to thrive in nutrient-poor soil, enriching the environment for other plants and contributing to the intricate web of life within the rainforest. .

The fight against pests and herbivores is a constant battle, and the Amazon's plant life has evolved a panoply of defense mechanisms. Physical deterrents, like the thorns of the passion fruit vine, Passiflora edulis, or the tough, fibrous leaves of the Amazonian bamboo, Guadua angustifolia, deter foraging animals. Chemical defenses, such as the toxic alkaloids produced by the Amazonian rubber tree, Hevea brasiliensis, and the pungent oils released by the Amazonian cinnamon, Cinnamomum amazonicum, serve to repel herbivores and protect against fungal infections. .

The Amazon's plants have even developed complex relationships with animals, co-evolving to facilitate pollination and seed dispersal. The vibrant, fragrant flowers of the Amazonian orchid, Cattleya trianae, entice pollinators like hummingbirds and butterflies, ensuring the survival of their species. The fleshy fruits of the Amazonian palm, Astrocaryum murumuru, provide a nutritious meal for animals like monkeys and parrots, who inadvertently disperse their seeds across the rainforest, contributing to the plant's expansion. .

The tapestry of life within the Amazon is a constant dance of adaptation and evolution. Plants, sculpted by the demands of their environment, exhibit an incredible range of strategies, showcasing the boundless creativity of nature. From the intricate design of their leaves to the complex mechanisms of their defense, the Amazon's plants are a testament to the power of natural selection, each one a masterpiece of adaptation in this vibrant and dynamic ecosystem. .

The Plants' Role in the Ecosystem: Producing Oxygen, Providing Food, and Supporting Life

The Plants: A Symphony of Colors and Textures.

The Amazon rainforest, a sprawling tapestry of emerald green, vibrant hues, and intricate textures, is a testament to the profound role plants play in the grand symphony of life. It's not just a visual spectacle; it's an intricate network of interdependence, where each plant, from the towering kapok tree to the delicate orchids clinging to its trunk, contributes to the delicate balance of the ecosystem. The very air we breathe, the food we consume, and the intricate web of life that surrounds us are all inextricably linked to the silent, yet potent, power of plants.

Imagine a world without the verdant canopies that filter sunlight, allowing the rainforest floor to be bathed in a dappled light. Without the intricate network of roots that hold the soil, preventing erosion and providing stability to the entire ecosystem. And imagine the stark silence of a world devoid of the buzzing of pollinators, drawn to the nectar-rich blossoms, or the rustling of leaves as animals seek shelter and sustenance. These are but a few of the myriad ways plants orchestrate the life within the Amazon, proving their role is not merely passive, but rather a central, driving force.

Oxygen, that invisible life-giving gas, is the very foundation of our existence. And plants, through the remarkable process of photosynthesis, are the masters of its production. Their leaves, a kaleidoscope of green, serve as solar panels, capturing the energy of sunlight. This captured energy powers the complex chemical reactions that transform carbon dioxide and water into sugars, providing the plant with nourishment and, as a byproduct, releasing oxygen into the atmosphere. The Amazon, with its vast expanse of lush vegetation, serves as a colossal oxygen factory, supplying not only its own teeming population but also contributing significantly to the global oxygen supply.

The Amazon's vibrant tapestry is not merely a visual delight; it's a veritable cornucopia of food, providing sustenance for a diverse array of life forms. The towering trees, with their fruits, nuts, and seeds, are a vital source of nourishment for countless creatures, from small insects to large mammals. The dense undergrowth, a mosaic of ferns, vines, and shrubs, offers a plethora of edible plants, each playing a unique role in the delicate food web. .

The plants within this ecosystem are not just a source of sustenance, they are also the architects of a complex network of interdependence. Each plant, in its unique way, provides a habitat, a refuge, or a food source for specific animals. For instance, the towering kapok tree, with its expansive branches, offers nesting sites for a variety of birds. The epiphytes, clinging to the branches, provide a home for insects and amphibians. The intricate network of vines, weaving through the canopy, serves as pathways for animals, providing access to hidden treasures and connecting different parts of the forest.

But the role of plants in the Amazon extends beyond the visible. Their roots, a complex subterranean network, not only anchor them to the soil but also play a crucial role in nutrient cycling and water management. The roots of trees, intertwined with the roots of other plants, form a vast, interconnected system that facilitates the transfer of water and nutrients, ensuring the ecosystem's vitality. The dense network of roots also helps stabilize the soil, preventing erosion and landslides, a crucial factor in the Amazon's ability to withstand heavy rainfall and maintain its ecological integrity.

The Amazon is not just a collection of individual plants; it is a complex, vibrant ecosystem where each plant plays a critical role. They provide the oxygen we breathe, the food we eat, and the habitats that support an incredibly diverse array of life. Their intricate relationship with the animals and the environment is a delicate dance, a symphony of colors and textures that speaks volumes about the profound impact of plants on the ecosystem, and ultimately, on the very fabric of life. .

Threats to Amazonian Plants: Habitat Loss and Climate Change

The Amazon rainforest, a verdant tapestry woven from countless species, stands as a testament to the extraordinary diversity of life on Earth. Within its emerald embrace, a symphony of colors and textures unfolds, painting a vivid portrait of nature's artistry. Yet, beneath this captivating facade, a silent struggle for survival plays out, fueled by the twin threats of habitat loss and climate change. .

The Amazon's flora, a vibrant mosaic of towering trees, vibrant flowers, and intricate ferns, faces an unprecedented assault on its very existence. The relentless march of deforestation, driven by agricultural expansion, illegal logging, and mining, disrupts the delicate balance of this ecosystem. Vast swaths of forest succumb to the relentless advance of human activity, leaving behind fragmented landscapes that are ill-suited to sustaining the rich tapestry of life.

As forests are cleared, the intricate network of relationships that bind plants and animals together unravels. The loss of habitat disrupts pollination processes, vital for the reproduction of many plant species. The disappearance of seed dispersal agents, such as birds and mammals, further jeopardizes the survival of plants reliant on these services. Without the protection of the canopy, seedlings struggle to thrive in the harsh, sun-baked conditions, leaving a barren landscape behind.

The specter of climate change looms large over the Amazon's future, casting a long shadow on its flora. Rising temperatures, altered rainfall patterns, and increased frequency of droughts disrupt the delicate balance that governs the ecosystem's equilibrium. Many plant species, adapted to specific climatic conditions, find themselves struggling to cope with the changing environment. Droughts, prolonged and severe, inflict stress on plants, making them vulnerable to disease and pests. As temperatures rise, the threat of wildfires escalates, leaving behind a scorched earth devoid of life.

The impact of climate change extends beyond the direct effects of altered weather patterns. Rising temperatures accelerate the rate of decomposition, releasing vast amounts of carbon dioxide into the atmosphere, further amplifying the greenhouse effect. This feedback loop exacerbates the already alarming trend of global warming, pushing the Amazon further into a perilous state. The delicate balance of the rainforest's carbon cycle is disrupted, shifting from a carbon sink to a carbon source, further accelerating the pace of climate change.

The Amazon's plant life, a cornerstone of this extraordinary ecosystem, stands at a critical juncture. The combined pressures of habitat loss and climate change threaten to unravel the very fabric of this biodiversity hotspot. The loss of plant species translates into a loss of ecosystem services, impacting not only the Amazon itself but also the global climate and human well-being. The Amazon's vast carbon storage capacity, a vital buffer against climate change, is steadily eroding, exacerbating the global climate crisis.

The intricate tapestry of the Amazon's flora, a testament to millions of years of evolution, is now facing a fight for survival. The urgency of the situation cannot be overstated. The fate of this extraordinary ecosystem, and the myriad life forms it supports, hangs precariously in the balance. The time for action is now. A concerted effort, encompassing policy changes, sustainable land management practices, and global collaboration, is needed to safeguard the

Amazon's biodiversity and preserve this vital life-support system for generations to come.

Chapter 13: The Future of the Amazon: Challenges and Opportunities

The Amazon's Importance to the World: A Global Resource and a Vital Ecosystem

The Amazon's Importance to the World: A Global Resource and a Vital Ecosystem.

The Amazon rainforest, a sprawling expanse of verdant life, is not merely a picturesque landscape; it is a critical cog in the Earth's intricate machinery, playing an indispensable role in regulating our planet's climate and sustaining a staggering biodiversity. Its importance transcends national borders, extending a lifeline to the entire world.

The Amazon acts as a colossal carbon sink, absorbing vast quantities of atmospheric carbon dioxide through its dense vegetation. This vital service mitigates the effects of climate change, slowing the rate of global warming and its associated consequences. Its sprawling forests act as a natural buffer against the relentless onslaught of climate change, offering a crucial reprieve from the escalating global temperature.

Beyond its role in climate regulation, the Amazon is a reservoir of unparalleled biodiversity. Its intricate ecosystems are home to an astonishing array of flora and fauna, harboring an estimated 10% of the world's known species. This biodiversity fuels the development of crucial medicines, agricultural resources, and even biofuels, providing a treasure trove of potential solutions for the world's growing needs. The Amazon, in essence, serves as a living laboratory, a fertile ground for scientific discovery and innovation, its secrets promising to unlock a wealth of knowledge and opportunities for the benefit of humankind.

The Amazon's influence extends far beyond its borders. The rivers that flow through the rainforest provide sustenance to millions of people downstream, while its abundant resources, from timber to rubber, have historically fueled global economies. This economic significance, however, comes with its own set of challenges, as unsustainable practices threaten the very ecosystem upon which these benefits rely.

The Amazon's vulnerability to human activities is a stark reminder of its delicate balance. Deforestation, driven by agricultural expansion, logging, and mining, is rapidly eroding the rainforest's ability to perform its vital functions. The consequences are dire: the loss of biodiversity, the disruption of vital hydrological cycles, and the release of vast amounts of stored carbon, accelerating the pace of climate change.

The challenge, therefore, lies in achieving a sustainable balance between utilizing the Amazon's resources and preserving its integrity. This calls for responsible governance, collaborative efforts between nations, and innovative solutions that foster economic development without sacrificing the environment. The Amazon is not merely a global resource; it is a shared responsibility, demanding a global response.

The future of the Amazon rests upon the shoulders of humanity, demanding a collective commitment to its preservation. This responsibility extends beyond governments and institutions, reaching every individual on the planet. By embracing sustainable practices, supporting conservation efforts, and demanding responsible resource management, we can safeguard the Amazon's vital role in our planet's future.

The fate of the Amazon is inextricably linked to the fate of the world. Its preservation is not just a matter of environmental concern; it is a matter of global survival. The Amazon, in its magnificent complexity, represents a crucial link in the intricate web of life on Earth, demanding our respect, protection, and unwavering commitment to its future.

Threats to the Amazon: Deforestation, Climate Change, and Pollution

The Amazon rainforest, a verdant tapestry of life encompassing a vast swathe of South America, stands as a global treasure trove of biodiversity. Its ecological significance, however, is increasingly threatened by a convergence of human activities – deforestation, climate change, and pollution. The future of this iconic ecosystem hangs precariously in the balance, demanding a concerted global effort to address these pressing challenges.

Deforestation, driven by agricultural expansion, logging, and mining, is a relentless assault on the Amazon's heart. The conversion of pristine rainforest into vast monocultures, especially for soy and cattle ranching, has led to a dramatic loss of habitat, impacting countless species that depend on the intricate web of life within the rainforest. This fragmentation of the rainforest disrupts vital ecological processes, hindering the movement of animals and the dispersal of seeds, further contributing to the decline of biodiversity. The devastating consequences of deforestation extend beyond the loss of habitat. The removal of trees, which act as carbon sinks, exacerbates climate change by releasing massive amounts of carbon dioxide into the atmosphere.

Climate change poses a multifaceted threat to the Amazon, further amplifying the effects of deforestation. Rising temperatures, altered precipitation patterns, and increased drought stress are pushing the rainforest towards a tipping point. Studies predict that the Amazon could transition from a lush rainforest to a dry savanna-like ecosystem, leading to a cascade of ecological and social consequences. This shift would not only decimate biodiversity but also impact global climate regulation, potentially exacerbating climate change. The intricate interplay between deforestation and climate change creates a dangerous feedback loop, making the Amazon increasingly vulnerable to both human and natural disturbances.

Pollution, a insidious threat emanating from various sources, adds another layer of complexity to the Amazon's challenges. The pervasive presence of pollutants like mercury, pesticides, and industrial waste contaminates the rainforest's ecosystems, impacting everything from aquatic life to the health of indigenous communities. Mining activities, particularly gold mining, often employ mercury, which leaches into rivers and accumulates in the food chain, posing severe health risks to both humans and wildlife. Agricultural practices, reliant on synthetic fertilizers and pesticides, contaminate waterways and soil, harming the intricate web of life that sustains the rainforest. The accumulation of pollutants in the Amazon's ecosystems creates a toxic legacy, threatening the health and well-being of its inhabitants.

While the threats to the Amazon are formidable, there are glimmering rays of hope in the form of innovative solutions and dedicated efforts aimed at preserving this invaluable ecosystem. Sustainable forest management practices, focused on responsible logging and reforestation, can help mitigate deforestation while providing economic benefits to local communities. Combating illegal activities like logging and mining requires robust law enforcement and international cooperation. The development and adoption of cleaner technologies in mining and agriculture can significantly reduce pollution levels, protecting both human health and the environment.

The conservation of the Amazon requires a multifaceted approach that integrates science, policy, and community engagement. Collaborative efforts between governments, NGOs, and local communities are essential for effective conservation. Investing in sustainable livelihoods for communities living within the Amazon's borders, fostering sustainable economic development, and promoting ecotourism can help ensure the long-term well-being of the rainforest and its inhabitants.

The Amazon, a vital part of the Earth's biosphere, stands as a testament to the interconnectedness of life on our planet. Its future hangs in the balance, demanding our collective action to protect this ecological wonder for generations to come. Only by confronting the challenges of deforestation, climate change, and pollution can we hope to secure a future where the Amazon continues to thrive, a vibrant tapestry of life pulsing with the rhythm of nature.

The Need for Conservation: Protecting a Precious Heritage

The Need for Conservation: Protecting a Precious Heritage.

The Amazon rainforest, a verdant tapestry woven from life in countless forms, stands as a testament to the Earth's biodiversity. Its rivers, teeming with fish, snake through a landscape of towering trees, their branches a haven for countless birds and mammals. Beneath the canopy, a world of vibrant insects, amphibians, and reptiles thrives, each playing a vital role in the delicate balance of this ecosystem. But this paradise, this jewel of the planet, is under threat. The forces of deforestation, driven by insatiable human demands, are pushing the Amazon to the brink, threatening not only its unique biodiversity but also the global climate and the well-being of millions of people.

The need for conservation in the Amazon is not a matter of choice, but a moral imperative. It is a commitment to safeguarding a legacy that transcends generations, a legacy of biological wonder and essential ecological services. The Amazon's intricate web of life provides us with clean air and water, regulates the global climate, and offers a treasure trove of medicinal plants and potential solutions to countless challenges facing humanity. Its preservation is not simply an act of environmentalism, but a vital step towards ensuring our own survival.

The Amazon's unique flora and fauna are a testament to its remarkable biodiversity. From the majestic jaguars prowling the dense undergrowth to the vibrant macaws soaring through the canopy, its inhabitants showcase a staggering array of adaptations and evolutionary marvels. The sheer diversity of plant life, from towering kapok trees to delicate orchids, reveals the immense complexity of this ecosystem. Each species, from the smallest insect to the largest mammal, plays a crucial role in maintaining the delicate balance that sustains this biological haven. .

The loss of this biodiversity is not just an aesthetic tragedy, it is a loss of essential ecosystem services. The Amazon's vast forests act as giant carbon sinks, absorbing vast amounts of carbon dioxide from the atmosphere and mitigating climate change. Its rich soil, teeming with microbes, contributes to the global nitrogen cycle, essential for plant growth. Its rivers, teeming with fish, provide sustenance for countless communities. The loss of these services would have devastating consequences for the planet and its inhabitants.

The threats to the Amazon are multifaceted. Deforestation, driven by agricultural expansion, logging, and mining, is relentlessly carving away at the rainforest. The insatiable demand for beef, soy, and other commodities is fueling the destruction of vast swaths of forest, displacing wildlife and altering the delicate balance of the ecosystem. Climate change, with its increasing temperatures and altered rainfall patterns, is adding to the pressure, making the Amazon more vulnerable to fire and drought. The combined effect of these threats is a slow but steady erosion of the rainforest, threatening to turn this vibrant ecosystem into a barren landscape.

The need for conservation in the Amazon is not simply about preserving its beauty, but about ensuring the survival of countless species and the well-being of generations to come. It is a call for a change in our relationship with the environment, a shift from exploitation to stewardship. It is a commitment to protecting a legacy that is not ours to destroy, but ours to cherish and pass on to future generations.

The Amazon's future hangs in the balance. The choices we make today will determine the fate of this vital ecosystem and its countless inhabitants. The path forward requires a concerted effort from governments, businesses, and individuals. We must prioritize sustainable development, promote responsible land use, and invest in conservation initiatives. We must recognize that the fate of the Amazon is inextricably linked to the fate of humanity. We must act now, for the sake of the planet and for the sake of ourselves.

The Amazon rainforest stands as a beacon of hope and a symbol of the Earth's resilience. Its preservation is a shared responsibility, a testament to our commitment to a sustainable future. In the face of mounting challenges, we must act with unwavering resolve to protect this precious heritage, to ensure that the Amazon's song continues to resonate through the ages. .

Opportunities for Sustainability: Working Towards a Balanced Future

Opportunities for Sustainability: Working Towards a Balanced Future.

The Amazon, a sprawling tapestry of emerald green, is a symphony of life, a throbbing heart of biodiversity. Its vastness holds within it a potential for a future that is both sustainable and prosperous, but the path to this future is paved with challenges. The rainforest, a vital resource for the planet and its inhabitants, faces unprecedented threats from deforestation, climate change, and unsustainable practices. Yet, amidst these challenges, lies a glimmer of hope - an opportunity to harness the Amazon's potential for a balanced future, a future where economic development walks hand in hand with environmental protection. .

The path towards a sustainable future in the Amazon begins with recognizing the interconnectedness of the region's ecosystems, its inhabitants, and the world beyond. The Amazon's intricate web of life, from the towering canopy trees to the smallest insects, is a delicate balance that sustains not only the region but also the global climate system. Understanding this interconnectedness is crucial for fostering responsible practices that respect the Amazon's delicate equilibrium. .

One of the key opportunities for sustainability in the Amazon lies in the realm of sustainable forestry. Traditional logging practices have often resulted in deforestation, leading to biodiversity loss and carbon emissions. However, with a shift towards responsible forestry, where harvesting is balanced with reforestation and sustainable practices, the Amazon's resources can be utilized for economic benefit without compromising its environmental integrity. This transition requires a collaboration between governments, communities, and businesses to develop and implement certified sustainable forest management practices, ensuring long-term ecological and economic viability.

Beyond timber, the Amazon offers a wealth of other potential resources for sustainable development. The region's rich biodiversity provides a vast reservoir of medicinal plants, many with untapped potential for pharmaceutical and biotechnological applications. Developing and utilizing these resources sustainably requires research, innovation, and ethical practices that ensure equitable access and benefit sharing. This endeavor can create economic opportunities while simultaneously safeguarding the region's biodiversity and cultural heritage.

Ecotourism presents another opportunity for sustainable development in the Amazon. By promoting responsible tourism practices that minimize environmental impact and prioritize the well-being of local communities, the region can attract visitors seeking authentic experiences while contributing to local economies. This approach requires careful planning, collaboration between tourism operators, and community involvement to ensure that tourism benefits the local population while respecting the environment.

Furthermore, the Amazon's vast natural resources, particularly its rivers, offer potential for clean and renewable energy generation. Hydropower, solar energy, and biomass energy can contribute to a low-carbon future, reducing the region's dependence on fossil fuels while mitigating climate change impacts. Harnessing these resources responsibly, considering the potential ecological consequences and ensuring community participation, is crucial for achieving a balanced future. .

Sustainable agriculture practices are vital for the future of the Amazon. Traditional slash-and-burn agriculture, often used for small-scale farming, has contributed significantly to deforestation. However, alternative practices such as agroforestry, where trees are integrated into agricultural systems, can provide both food security and environmental benefits. Promoting sustainable agricultural practices requires knowledge sharing, technical support, and access to resources for small-scale farmers, empowering them to participate in a sustainable future. .

The fight against deforestation is an integral part of achieving sustainability in the Amazon. Developing and implementing effective mechanisms for monitoring and controlling deforestation is essential, using technology and collaborative efforts to protect the rainforest's integrity. Combating illegal logging and land grabbing, promoting alternative livelihoods for those reliant on forest resources, and strengthening legal frameworks are critical steps towards a sustainable future. .

The Amazon's future is not just about its own ecosystems, but also about its human inhabitants. The indigenous communities of the Amazon have a deep understanding of the region's resources and a rich cultural heritage that is inextricably linked to the rainforest. Recognizing their rights and knowledge is essential for achieving a sustainable future. Involving them in decision-making processes, respecting their traditional practices, and ensuring equitable benefit sharing are fundamental principles for a just and sustainable future for the Amazon.

A sustainable future for the Amazon requires a shift in perspective, moving away from the view of the region as a resource bank to one that recognizes its intrinsic value. This shift involves recognizing the interconnectedness of nature, valuing its ecological services, and acknowledging the rights of its inhabitants. It necessitates collaboration between governments, communities, businesses, and researchers, fostering a shared vision for a balanced future.

The path to a sustainable Amazon is not without its challenges. The complex social, economic, and political dynamics of the region require careful navigation. Balancing the needs of development with the demands of environmental protection necessitates innovative approaches, inclusive decision-making, and a commitment to long-term sustainability.

Yet, despite the challenges, the future of the Amazon holds immense potential. By embracing a vision of sustainability, by harnessing the region's vast resources responsibly, and by acknowledging the rights and knowledge of its inhabitants, we can pave the way for a balanced future, a future where the Amazon flourishes, not only for its inhabitants but for the planet as a whole. The

Amazon's future is a testament to the power of collaboration, the resilience of nature, and the potential for human ingenuity to create a world where both humanity and the environment can thrive.

Printed in Great Britain
by Amazon